Springer Complexity

Springer Complexity is an interdisciplinary program publishing the best research and academic-level teaching on both fundamental and applied aspects of complex systems—cutting across all traditional disciplines of the natural and life sciences, engineering, economics, medicine, neuroscience, social and computer science.

Complex Systems are systems that comprise many interacting parts with the ability to generate a new quality of macroscopic collective behavior the manifestations of which are the spontaneous formation of distinctive temporal, spatial or functional structures. Models of such systems can be successfully mapped onto quite diverse "real-life" situations like the climate, the coherent emission of light from lasers, chemical reaction-diffusion systems, biological cellular networks, the dynamics of stock markets and of the internet, earthquake statistics and prediction, freeway traffic, the human brain, or the formation of opinions in social systems, to name just some of the popular applications.

Although their scope and methodologies overlap somewhat, one can distinguish the following main concepts and tools: self-organization, nonlinear dynamics, synergetics, turbulence, dynamical systems, catastrophes, instabilities, stochastic processes, chaos, graphs and networks, cellular automata, adaptive systems, genetic algorithms and computational intelligence.

The three major book publication platforms of the Springer Complexity program are the monograph series "Understanding Complex Systems" focusing on the various applications of complexity, the "Springer Series in Synergetics", which is devoted to the quantitative theoretical and methodological foundations, and the "Springer Briefs in Complexity" which are concise and topical working reports, case studies, surveys, essays and lecture notes of relevance to the field. In addition to the books in these two core series, the program also incorporates individual titles ranging from textbooks to major reference works.

Indexed by SCOPUS, INSPEC, zbMATH, SCImago.

Understanding Complex Systems

Rafael Martínez-Guerra · Juan Pablo Flores-Flores

An Approach to Multi-agent Systems as a Generalized Multi-synchronization Problem

Springer

Rafael Martínez-Guerra
Automatic Control
CINVESTAV-IPN
Mexico City, Mexico

Juan Pablo Flores-Flores
Automatic Control
CINVESTAV-IPN
Mexico City, Mexico

ISSN 1860-0832 ISSN 1860-0840 (electronic)
Understanding Complex Systems
ISBN 978-3-031-22671-7 ISBN 978-3-031-22669-4 (eBook)
https://doi.org/10.1007/978-3-031-22669-4

This Springer imprint is published by the registered company Springer Nature Switzerland AG
The registered company address is: Gewerbestrasse 11, 6330 Cham, Switzerland

In memory of my father,
Carlos Martínez Rosales.
To my wife and sons,
Marilen, Rafael and Juan Carlos.
To my mother and brothers,
Virginia, Victor, Arturo, Carlos,
Javier and Marisela.

Rafael Martínez-Guerra

To my family,
Eladio, Candelaria, Edy, Abril, Mario,
Amaia, Akane and Najmeh.

Juan Pablo Flores-Flores

Preface

This book aims to expose the reader to the latest advances in chaotic synchronization and in particular, multi-synchronization of integer and fractional order. The content is ideal for graduate students and researchers in the control theory, nonlinear systems, multi-agent systems and synchronization fields. Throughout this book, diverse analytic, algebraic, differential-algebraic, geometric and asymptotic concepts are used to develop several results and algorithms for controlling and observing a wide variety of nonlinear systems in a master–slave configuration, which can be represented either as a state-space model or an input–output equation. The observations made in nature, in particular in multiple coupled oscillators, have inspired and motivated the study of the synchronization phenomenon and the corresponding extensions to more complex cases. These studies have led to important technological applications in fields such as secure communications, encryption algorithms, distributed computing, sensor networks and so on. Nowadays, the synchronization of multiple coupled systems, known as multi-synchronization, can be understood as a multi-agent system (MAS) problem, i.e., in a network of dynamic systems, where each of these is known as an agent, the information is shared in the network such that every agent follows a trajectory, accordingly to the interaction or interactions that each system has. To model the interactions on the network, graph theory can be used to describe the topology of the MAS. In a common topology, there exists a leader, this is, among the dynamic systems of the network, a system known as master guides the rest that are known as slaves. Thus, by means of a coupling signal between the slaves and the master, the trajectory of every slave system (also known as followers) follows the trajectory of the leader. In the literature, the last problem is known as the consensus following the leader. Another possibility is a network that has no leader, such that the interactions between the systems of the MAS can take any form. In this scheme, the trajectories of all the agents follow a given path that might be given by an external system, this is, a system that does not belong to the network. In a multi-synchronization problem, increasing the complexity of the interactions between the agents in the network is a key aspect, such as considering intermittent interactions. Moreover, in reality, the systems of the MAS are rarely identical. To study these more complex scenarios, it is possible to consider networks of heterogeneous chaotic systems, that is to say, to study the multi-synchronization problem in a generalized sense, or in other words, to study the generalized multi-synchronization (GMS) problem. In such scenario, the main issue is to prove the existence of differentially continuous mappings that relate the trajectories of the master system with the ones of each slave system. This topography is given by a direct invariant in time graph (unidirectional couplings) which is not strongly connected, this is, a directed tree of coverage with a single leader as a root. Although this graph is somewhat trivial, the GMS problem is not trivial, since it still demands to find the mappings or transformations between master and slaves and to verify the stability. The last is the main topic that this book addresses, and several mathematical tools, including differential algebra and fractional calculus tools, are presented to solve it. This book is organized as follows: In Chap. 1, a general overview of the synchronization problem is discussed. It is seen that from the point of view of control theory, synchronization of chaotic systems can be seen as a trajectory tracking problem. Finally, it is shown how the consensus problem can be traduced to a multi-synchronization problem. In Chap. 2, the generalized synchronization (GS) concept, briefly described in Chap. 1, is explored in detail. The relevance of GS when non-identical systems need to be synchronized is seen. Moreover, some traditional approaches for tackling this problem are briefly introduced to then, describing the differential algebra approach. Chapter 3 is dedicated to an alternative way of seeing synchronization, this is, a state estimation problem. Based on some algebraic properties, a state observer is presented. Moreover, the proposed observer is implemented in a current real-world problem, that is to say, the infected populations during the COVID-19 pandemic are studied by considering an epidemiological model. Chapter 4 describes how the consensus problem can be traduced to a generalized multi-synchronization problem, where there is only one leader or master system and several followers/slaves. This topic is extended in Chap. 5, this is, now is considered that the followers interact between them, not just with the master. In this chapter, the convergence analysis is conducted by considering the input-to-state convergence definition. In Chap. 6, partial differential equation systems are the main concern. Here, both generalized synchronization and generalized multi-synchronization for this class of systems are explored. It is worth mentioning that, unlike in previous chapters, here the multi-synchronization problem considers families of masters and slaves. Finally, Chap. 7 is dedicated to systems of fractional order. Therefore, several new concepts and definitions from fractional calculus are introduced. This chapter also considers PDE systems whose time derivative is of fractional order. Every single chapter is supported by a series of numerical examples that were selected to illustrate the possible applications to real-world problems. The authors have attempted to write in such a way that this book can be read by an audience as wide as possible.

Mexico City, Mexico

Rafael Martínez-Guerra
Juan Pablo Flores-Flores

Contents

Notations and Abbreviations

$\doteq$	Defined as
$\approx$	Approximately equal to
$:=$	Defined as
$\neq$	Not equal to
$<$ $(>)$	Less (greater than)
$\leq$ $(\geq)$	Less (greater) than or equal to
$\forall$	For all
$\in$	Belongs to
$\subset$	Subset of
$\cup$	Sets union
$\{\cdot\}$	Set
$\rightarrow$	Tends to
$\mathbb{R}$	Set of real numbers
$\mathbb{Z}$	Set of integer numbers
$\mathbb{Z}^+$	Set of integer and positive numbers
$\mathbb{R}^{n \times m}$	Set of all $n \times m$ matrices with elements in $\mathbb{R}$
A^T	Transpose of matrix A
v^T	Transpose of vector v
A^{-1}	Inverse of matrix A
$A > 0$	Positive definite matrix A
$\Vert\cdot\Vert$	Euclidean norm
$\vert\cdot\vert$	Absolute values
$\square$	End of proof
$F : S_1 \rightarrow S_2$	A function f mapping a set S_1 into a set S_2
$\mathrm{K} < \cdot >$	Differential field K
$\mathrm{K} < \cdot > /\mathrm{L} < \cdot >$	K is a differential field extension of L $(L \subset K)$
$\mathscr{D}^{r\alpha}$	Sequential fractional-order derivative operator
$\otimes$	Kronecker product
$\dot{y}$	Time derivative of y
$y^{(n)}$	n-th time derivative of y
$\dot{y} = \frac{\partial y}{\partial t}$	Partial derivative of y with respect to time

$y_x = \frac{\partial y}{\partial x}$	Partial derivative of y with respect to x
$y_{xx} = \frac{\partial^2 y}{\partial x^2}$	Second partial derivative of y with respect to x
AOC	Algebraic observability condition
CS	Complete synchronization
FGMS	Fractional generalized multi-synchronization
FGS	Fractional generalized synchronization
GMS	Generalized multi-synchronization
GOCF	Generalized observability canonical form
GS	Generalized synchronization
MAS	Multi-agent system
MFGOCF	Multi-output fractional generalized observability canonical form
MGOCF	Multi-output generalized observability canonical form
PI	Proportional-integral
PV	Picard–Vessiot

List of Figures

Chapter 1
An Overview of Chaos Synchronization

Most of the literature on control systems focuses on mitigating and exploiting some of the intrinsic properties of dynamical systems. Therefore, to grasp the essence of the control point of view, we need to see dynamical systems as sets of differential equations with additional features (e.g., external noise, complex interactions, unmodeled dynamics, parameter uncertainties, etc.). Nonetheless, by considering all these features, control problems could be restricted to stochastic and robust control approaches. Therefore, to understand those, we first need to tackle the "easy ones", for which pure nonlinear differential equations appear (for an insightful discussion about this, see [39, 70]). One of the most common topics in control theory is the trajectory tracking problem. In this, given two or more dynamical systems, the idea is to follow a specific state trajectory or path. In general, the path to go after is predefined; however, it might be the case that such trajectory is generated by another dynamical system. In such a case, what we describe is known as synchronization.

The simplest form of *synchronization* between two dynamical systems happens when the state trajectory of one of them follows the other due to a external input or the existence of a coupling signal between them. This simple idea, which is ubiquitous, has been an interesting and inspiring area for researchers from different disciplines (biology, chemistry, physics, mathematics, astronomy, sociology, engineering and technology, among others) for more than 350 years.[1] In applied mathematics, and particularly in the interdisciplinary area of automatic control, these problems are extremely interesting and increasingly studied.

[1] This phenomenon was first reported by Christian Huygens in a letter to his father in 1665. In that letter, Huygens said that "sympathy" (synchronization) occurs between two "identical" maritime pendulum clocks (dynamical systems) when these are hung on the same wall (coupling) at some distance [24].

R. Martínez-Guerra and J. P. Flores-Flores, *An Approach to Multi-agent Systems as a Generalized Multi-synchronization Problem*, Understanding Complex Systems,
https://doi.org/10.1007/978-3-031-22669-4_1

1.1 Chaos Synchronization

The study of synchronization problem is broad and in general, is centered on the examination of *coupled oscillators*. The main reasons are that these systems are of (quasi) periodic behavior and are characterized by a strong influence of one over the other. That is to say, due to their influence, commonly given by a coupling signal, it is possible the existence of similar behavior. On the other hand, it is well known that coupled oscillators can pass between successive regimes of periodic and chaotic behavior, depending on their parameters [29].

Even synchronization occurs for periodical systems, the phenomenon is far more interesting when chaos is involved. A system that exhibits chaos or chaotic system is a deterministic system defined by a set of nonlinear equations $\dot{x} = f(x), x = x(t)$ that shows an apparent random behavior and satisfies the following properties:

(1) Bounded dynamics, that is to say, for the map $f : \mathscr{X} \to \mathscr{X}$, where $\mathscr{X}$ is a closed invariant set, there exists an attractive subset $\mathscr{C}$ such that for any $t \geq 0$ and $x \in \mathscr{X}$, we have that $f(x) \in \mathscr{X}$. Or equivalently, there exist a real positive number M such that $\|f(x)\| \leq M$ for any $x \in \mathscr{X}$.
(2) Sensitive dependence on initial conditions, this means that for some $\delta > 0$, any $x \in \mathscr{X}$ and any neighborhood $\mathscr{N}_\varepsilon(x) = (x - \varepsilon, x + \varepsilon)$ of x, there exits $\bar{x} \in \mathscr{N}_\varepsilon(x)$ and an integer $k > 0$ such that $\|f^k(x) - f^k(\bar{x})\| > \delta$ holds.

In principle, the behavior of a chaotic system is predictable; however, due to their sensitivity to initial conditions, in the practice, these are considered to be unpredictable. Some examples of chaotic systems are the systems of Lorenz, Rössler, Chua and Colpitts system [27, 61]. Since these systems have many possible applications, it is worth asking: Is it possible to force two deterministic systems with unpredictable aperiodic behavior and bounded trajectories that exponentially diverge to follow the same trajectory? As incredibly as it sounds, the answer to this question is yes and still fascinates the scientific community since its introduction by Louis M. Pecora and Thomas L. Carroll in 1990.[2]

When studying synchronization three main concerns arise: (1) How do systems interact with their neighbors? Therefore, it is important to assume some kind of configuration, such as a unidirectional or bidirectional coupling. (2) How do systems share their information about their complete or partial state? This is, the coupling is linear or nonlinear. And (3) how different or similar are the linear or nonlinear systems involved? This might refer to the system parameters or the differential equations. These questions must be answered before attempting to solve a synchronization problem.[3]

---

[2] Before 1990, synchronization and chaotic behavior were associated with an undesirable phenomenon with no useful engineering application [60]. Hence, chaos was studied as an outcome of the loss of synchronization and was mainly popular only among physicists. However, nowadays this phenomenon has been successfully implemented in active research areas such as secure communications [12, 13, 63, 65], cryptography [32], etc.

[3] The synchronization phenomenon has been widely studied. In literature, we can find several types of synchronization. The most common ones are: complete synchronization, generalized synchro-

1.2 Synchronous State

The problems associated with *synchronization of chaotic systems* are closely related to the stability of two sets, which define the geometry of this behavior [48], namely[4]:

- *Synchronization manifold*: It is the set of coordinates or trajectories of coupled systems that define the *synchronous state* (hyperplane). System trajectories are restricted to this set when synchronization exists.
- *Transverse manifold*: It is the space orthogonal to the synchronization manifold. It contains the set of zero coordinates or trajectories when motion is restricted to the synchronization manifold.

The *synchronous state* corresponds to chaotically behaved identical trajectories due to the coupling between the chaotic systems that generate them. That is, the synchronous state is a *chaotic attractor* within the synchronization manifold. An attractor [61] is a closed invariant set $\mathscr{C}$ such that fulfills the following conditions:

(1) Any trajectory $x \in \mathscr{C}$ keeps within $\mathscr{C}$ as $t \to \infty$.
(2) $\mathscr{C}$ attracts an open set $\mathscr{U}$ of initial conditions, where $\mathscr{C} \subset \mathscr{U}$. If $x(0) \in \mathscr{U}$, then the distance between $x(t)$ and $\mathscr{C} \to 0$ as $t \to \infty$. The largest set $\mathscr{U}$ is known as the *basin of attraction* of $\mathscr{C}$.
(3) $\mathscr{C}$ is minimal, there is no smaller proper subset of $\mathscr{C}$ which fulfills first two conditions.
(4) If the attractor exhibits *sensitive dependence on initial conditions* it is called chaotic attractor.

It is important to emphasize that transversal and synchronization manifolds can be associated with simpler concepts such as the synchronization error (or tracking error) and the origin of the synchronization error, respectively. Consequently, it is understood that the zero point within the transverse manifold of synchronization is a fixed point of such manifold, or equivalently, the origin is an equilibrium point of the synchronization error dynamics. With the following example, we attempt to illustrate these ideas.

---

nization, projective synchronization, phase synchronization, delayed synchronization and observer-based synchronization.

[4] These problems are equivalent to those formulated for synchronization of (quasi) periodic oscillators, limit cycles, etc. Later, its extension has been a case of study for those systems "who apparently defy synchronization" [44], i.e., chaotic systems.

1.2.1 Example of Chaos Synchronization

Let us consider two identical Lorenz chaotic systems. These are given by

$$\begin{aligned}\dot{x}_1 &= \sigma(y_1 - x_1),\\ \dot{y}_1 &= -x_1 z_1 + r x_1 - y_1,\\ \dot{z}_1 &= x_1 y_1 - b z_1\end{aligned}$$

and

$$\begin{aligned}\dot{x}_2 &= \sigma(y_2 - x_2),\\ \dot{y}_2 &= -x_2 z_2 + r x_2 - y_2,\\ \dot{z}_2 &= x_2 y_2 - b z_2\end{aligned}$$

with their respective initial conditions (different from each other) and identical parameters $\sigma, r, b > 0$. The trajectories $x_m = (x_1, y_1, z_1)^T$ and $x_s = (x_2, y_2, z_2)$ of these systems will diverge one from another over time due to the sensitivity to initial conditions (see Fig. 1.1).

In control theory, the so-called *master–slave synchronization scheme* refers to a configuration where one system receives information from the other, such that the state trajectories are influenced by the incoming signal. For convenience, from now on we will refer to the first system as *master system* and to the second one as *slave system*.

Back to our example, under the master–slave configuration[5] (Fig. 1.2), now we consider that master and slave systems are coupled by the first state, this is, the state x_2 of the slave is substituted by the state x_1 of the master in the corresponding dynamic equations of y_2 and x_2 as follows

$$\begin{aligned}\dot{x}_1 &= \sigma(y_1 - x_1),\\ \dot{y}_1 &= -x_1 z_1 + r x_1 - y_1,\\ \dot{z}_1 &= x_1 y_1 - b z_1\end{aligned}$$

and

$$\begin{aligned}\dot{x}_2 &= \sigma(y_2 - x_2),\\ \dot{y}_2 &= -x_1 z_2 + r x_1 - y_2,\\ \dot{z}_2 &= x_1 y_2 - b z_2\end{aligned}$$

[5] In some texts, the terminology transmitter–receiver is used instead of master–slave. The first term was adopted and presented in Kevin Cuomo's work about coupled circuits [12].

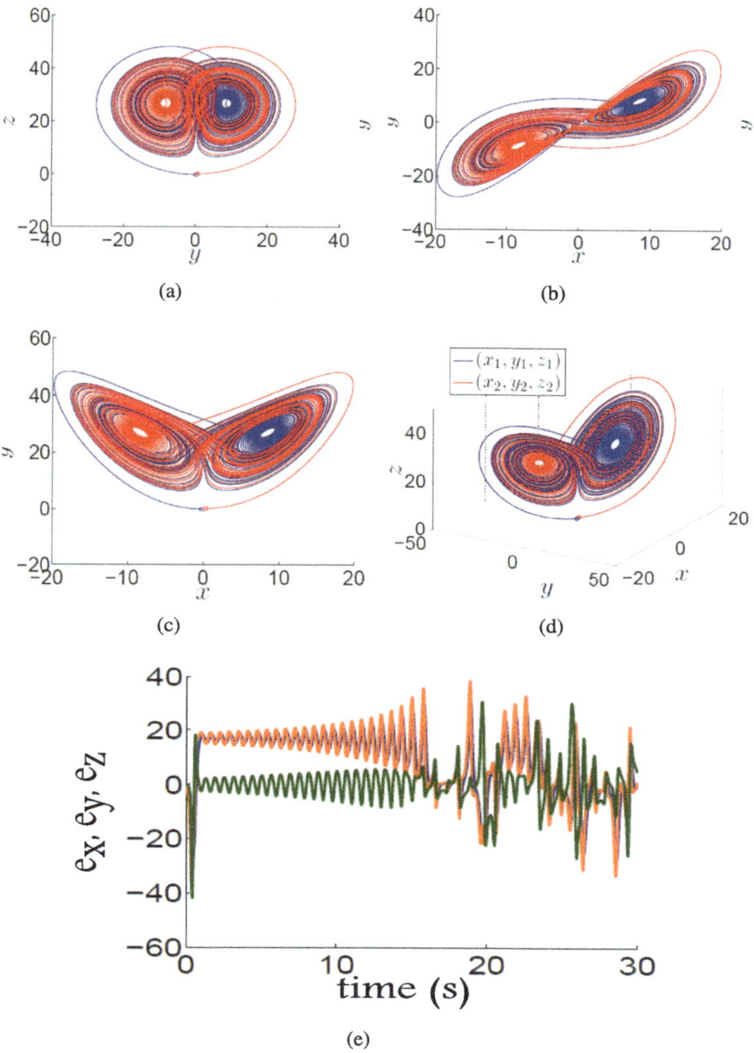

Fig. 1.1 Chaotic systems are highly sensitive to initial conditions. Consider two decoupled identical Lorenz chaotic systems with parameters $\sigma = 10$, $r = 8/3$, $b = 28$ and initial conditions $x_m(0) = (x_1(0), y_1(0), z_1(0))^T = (-0.2, 0.1, -0.1)^T$ and $x_s(0) = (x_2(0), y_2(0), z_2(0))^T = (0.3, 0.5, 0.2)^T$, respectively. The trajectories of these systems, initially close to each other, separate exponentially fast and evolve into different attractors, as shown in (**d**) and the projections over the different planes (**a**), (**b**), (**c**). The difference between trajectories $e_x(t) = x_1(t) - x_2(t)$, $e_y(t) = y_1(t) - y_2(t)$ and $e_z(t) = z_1(t) - z_2(t)$ clearly does not converge to zero when $t \to \infty$ (**e**)

Fig. 1.2 Master–slave synchronization scheme. The nodes represent the dynamics $\dot{x}_m = f_m(x_m)$ and $\dot{x}_s = f_s(x_s, u(x_m))$ associated with the master and slave systems, respectively. The arrow indicates the direction in which information flows. One can notice that the trajectory of the slave system x_s is influenced by an input that depends on the master system x_m

From the coupling between systems, intuitively it is understood that a chaotic signal, that behaves similarly to the one that replaces, will affect the dynamics of the slave system such that eventually the slave system will behave in the same way.[6] Using different conditions, both in the slave and in the master, it is observed that the trajectories of the slave x_2, y_2 and z_2 converge to the trajectories of the master x_1, y_1 and z_1, respectively. Equivalently, $|x_1 - x_2| \to 0$, $|y_1 - y_2| \to 0$ and $|z_1 - z_2| \to 0$ as $t \to \infty$ (see Fig. 1.3).

Now, let us define $x = (x_1, x_2)$, $y = (y_1, y_2)$ and $z = (z_1, z_2)$. From this, we conclude that the (trivial) synchronization manifold and the transversal manifold are described by

$$\mathscr{M} = \{(x, y, z) | x_1 = x_2, y_1 = y_2, z_1 = z_2\}$$

and

$$\mathscr{M}_\perp = \left\{(e_x, e_y, e_z) | e_x = x_1 - x_2, e_y = y_1 - y_2, e_z = z_1 - z_2\right\},$$

respectively.

1.3 Stability and Synchronization

From the perspective of automatic control, the study of synchronization lies in testing the stability of the synchronization manifold (if it exists) or the transversal manifold. It is not hard to notice that the stability of both manifolds is equivalent in the sense that one implies the other and vice versa. In the previous example, when systems are coupled, the synchronization occurs because the synchronization manifold is stable. To observe this, let us define synchronization errors e_x, e_y and e_z as in $\mathscr{M}_\perp$. Then, the zero point of the transversal manifold must be a fixed point within that manifold. This leads us to the fact that the origin is a point of stable equilibrium of the dynamics associated with e_x, e_y and e_z. From the dynamics of the synchronization error, we have

6 To control the chaos with chaos was the "cornerstone" for the progress made by Pecora in synchronizing chaos [60]: "I need to drive chaos with chaos—I need to drive the receiver with a signal that comes from the same kind of system".

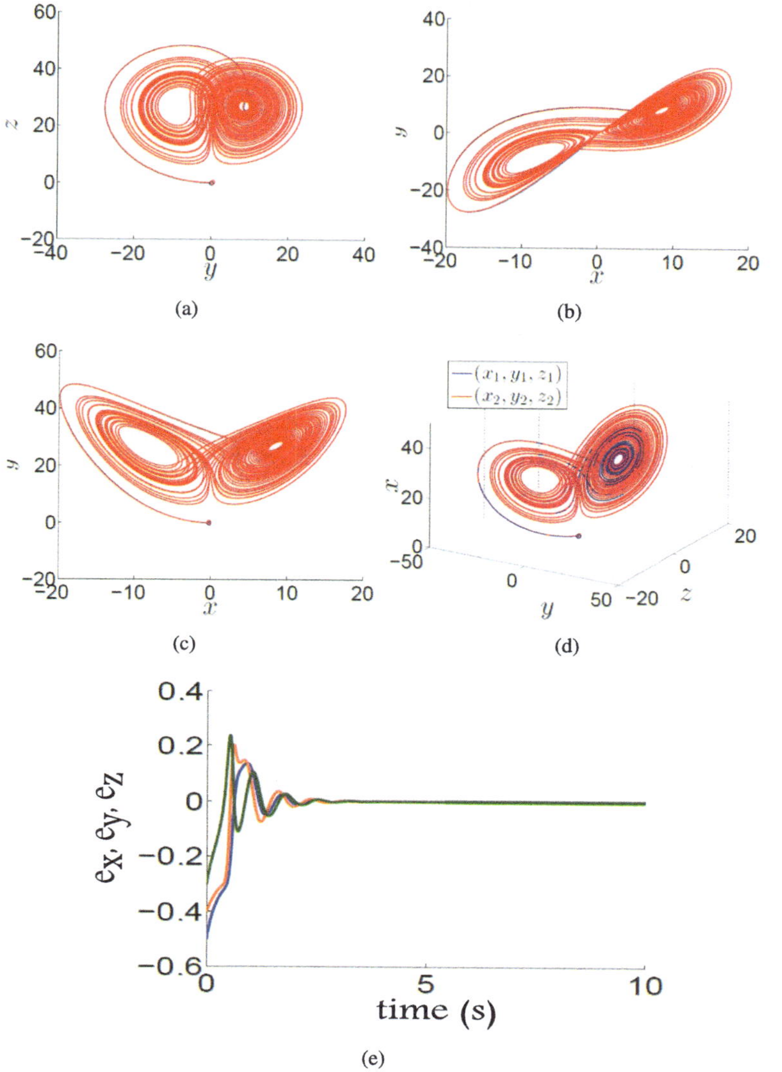

Fig. 1.3 Synchronization of identical Lorenz systems, coupled by x_1, with parameters $\sigma = 10$, $r = 8/3$, $b = 28$ and initial conditions $x_m(0) = (x_1(0), y_1(0), z_1(0))^T = (-0.2, 0.1, -0.1)^T$ and $x_s(0) = (x_2(0), y_2(0), z_2(0))^T = (0.3, 0.5, 0.2)^T$ respectively. Due to the replacement of x_2 by x_1, trajectories are restricted eventually to the synchronization manifold. In (**a**), (**b**) and (**c**) the projections of the synchronization manifold $x_1 = x_2$, $y_1 = y_2$ and $z_1 = z_2$ are observed. When systems evolve in the same chaotic attractor (**d**), the difference between trajectories $e_x(t) = x_1(t) - x_2(t) \to 0$, $e_y(t) = y_1(t) - y_2(t) \to 0$ and $e_z(t) = z_1(t) - z_2(t) \to 0$ as $t \to \infty$ (**e**)

$$\dot{e}_x = \sigma(e_y - e_x),$$
$$\dot{e}_y = -e_y - x_1 e_z,$$
$$\dot{e}_z = x_1 e_y - b e_z. \tag{1.1}$$

We can talk about stability if e_x, e_y and e_z tend to zero as $t \to \infty$. This is clear if we analyze the stability of the equilibrium point of system (1.1). This dynamic clearly depends on x_1; however, it is possible to obtain a *Lyapunov candidate function* and conclude about the stability of the synchronization error $e = (e_x, e_y, e_z)^T$. Let be $V(e) = \frac{1}{2\sigma}e_x^2 + \frac{1}{2}(e_y^2 + e_z^2)$, calculating the derivative of $V(e)$ along the trajectories of e we have that $\dot{V}(e) = -(e_x^2 - e_x e_y) - e_y^2 - b e_z^2 = -\left(e_x - \frac{1}{2}e_y\right)^2 - \frac{3}{4}e_y^2 - b e_z^2 \leq 0$ with $\dot{V}(e) = 0$ when $V(e) = 0$, therefore $e = 0$ is global, asymptotically stable. Note that the terms that depend on x_1 disappear after the derivation of $V(e)$. In consequence, as $\dot{V}(e) < 0$, it can be immediately concluded that the synchronization manifold is asymptotically stable.

For most of the cases, the previous method is conclusive [12, 13]. However, it is important to note that even when the systems are identical, the dynamics of the synchronization error depend on the states of both systems. Therefore, proposing a Lyapunov function is not simple. A more general solution is to study the stability of the variational equation associated with the transversal manifold (Taylor expansion of the vector field of the synchronization error, i.e., the Jacobian of the synchronization error dynamics). For (1.1), this equation is

$$\dot{e} = \begin{pmatrix} -\sigma & \sigma & 0 \\ 0 & -1 & -x_1 \\ 0 & x_1 & -b \end{pmatrix} e$$

where the minimum necessary condition for stability [48] of the synchronization error depends on the negativity of the (conditional) exponents of Lyapunov of this variational equation. Due to the structure of the dynamic equation of the synchronization error, when studying the stability of the synchronization or transversal manifolds is mainly considered a variational approach or, at best, Lyapunov's stability approach.

1.4 Synchronous State Problems

In practice, the negativity of the exponents of Lyapunov is not sufficient for the stability of the synchronous state. This is due to the maximum conditional exponent of Lyapunov which is almost independent of the attraction basin of the attractor [48]. For example, there are invariant sets (e.g., equilibrium points, periodic orbits, invariant manifolds, etc.) inside the chaotic attractor (in the synchronization manifold) whose maximum conditional exponent of Lyapunov is positive, even when the maximum exponent of Lyapunov of the attractor is negative. Thus, trajectories that approach the synchronization manifold are abruptly and/or intermittently repelled by these invariant sets [6, 9, 19, 20, 41, 48, 49]. These problems are well understood and can be solved appropriately using a powerful tool such as "Master Stability Function" (MSF) in the case of synchronization of (almost) identical systems [8, 20, 46, 62].

1.5 Generalized Synchronization and Associated Problems

The existence of a trivial and invariant manifold is a well-known fact in the problems of identical coupled oscillators [48]. In this particular case, variational methods such as MSF describe this phenomenon very well [8, 22, 46, 56, 62, 70]. However, real physical systems commonly involve groups of coupled systems with different dynamical structures, and therefore, these play an important role in the synchronization field [15, 53, 59]. In these cases, the stability of the synchronization error or the synchronization manifold is important and more challenging than in those with strictly identical systems. Consider the following Rössler and Lorenz systems in a master–slave configuration with diffusive coupling in the first component, i.e.,

$$\begin{aligned}\dot{x}_1 &= -(y_1 + z_1),\\ \dot{y}_1 &= x_1 + 0.2y_1,\\ \dot{z}_1 &= 0.2 + z_1(x_1 - \mu),\end{aligned}$$

and

$$\begin{aligned}\dot{x}_2 &= \sigma(y_2 - x_2) - g(x_2 - x_1),\\ \dot{y}_2 &= -x_2z_2 + rx_2 - y_2,\\ \dot{z}_2 &= x_2y_2 - bz_2,\end{aligned}$$

with parameters $\sigma, r, b, \mu > 0$, with coupling force $g \geq 0$. Consider the following situations: when the systems are decoupled (i.e., $g = 0$), it is clear that the trajectories of both systems evolve in different attractors (see Fig. 1.4). On the other hand, when the systems are coupled (i.e., $g > 0$), note that the trajectories of the slave x_2, y_2 and z_2 do not follow the master trajectories x_1, y_1 and z_1, that is, the asymptotic behavior of the synchronization error does not exist (as seen in Fig. 1.5). It could seem logical that increasing the value of g implies that the systems reach a synchronous state; however, this is not that simple.

At this point, as obtaining a dynamic equation of the synchronization error and analyzing its stability is not trivial, the following questions arise: Is there any relationship between the trajectories indicating that these systems are synchronized? and moreover, is it possible to determine the synchronous state of these systems? It is evident that the affirmation of these questions is irresistible given the possible applications, especially those suggested by Pecora and Carroll [44].

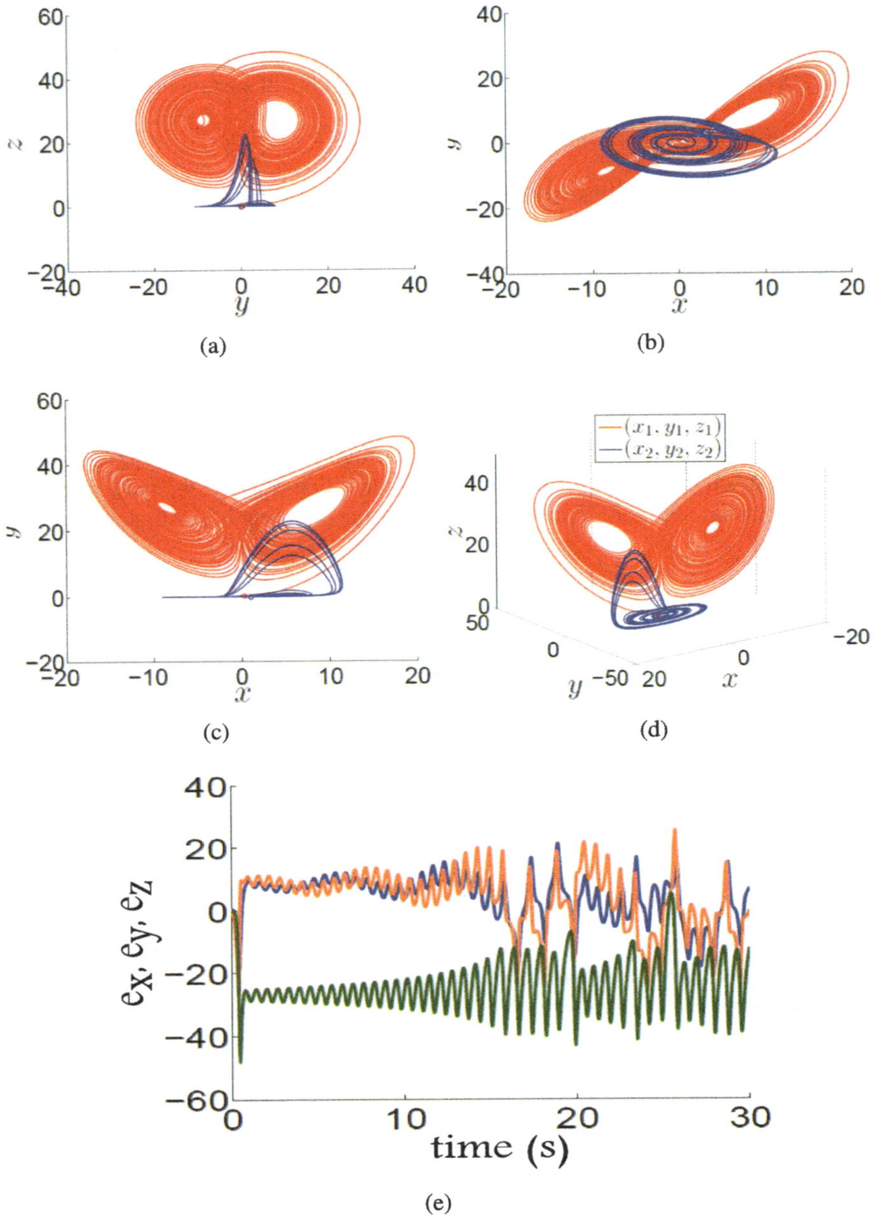

Fig. 1.4 Rössler and Lorenz systems decoupled with parameters $\mu = 5.7$, $\sigma = 10$, $r = 8/3$, $b = 28$ and initial conditions $x_m(0) = (x_1(0), y_1(0), z_1(0))^T = (1, 0.1, -0.1)^T$ and $x_s(0) = (x_2(0), y_2(0), z_2(0))^T = (0.3, 0.5, 0 - 2)^T$, respectively. The projections of the chaotic attractors (**a**), (**b**) and (**c**) show their evolution in different attractors (**d**). Clearly, the difference between the trajectories (**e**) of both systems $e_x(t) = x_1(t) - x_2(t)$, $e_y(t) = y_1(t) - y_2(t)$ and $e_z(t) = z_1(t) - z_2(t)$ does not converge to zero as $t \to \infty$

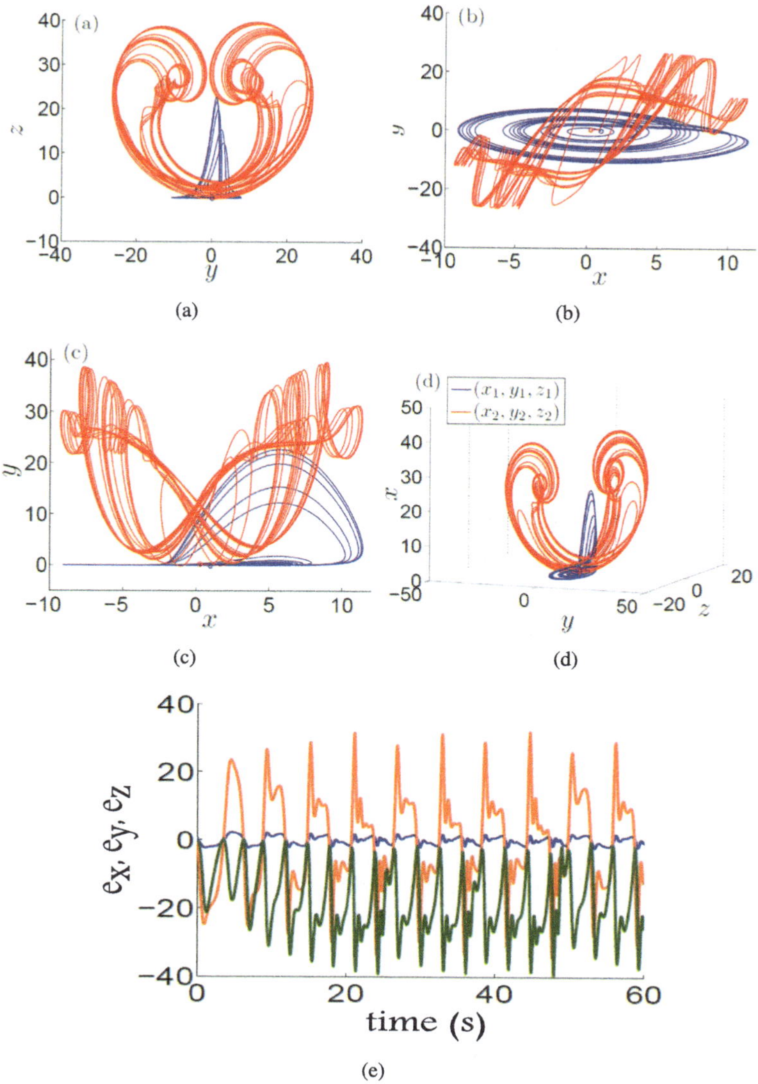

Fig. 1.5 Rössler and Lorenz systems coupled with parameters $\mu = 5.7$, $\sigma = 10$, $r = 8/3$, $b = 28$, $g = 100$ and initial conditions $x_m(0) = (x_1(0), y_1(0), z_1(0))^T = (1, 0.1, -0.1)^T$ and $x_s(0) = (x_2(0), y_2(0), z_2(0))^T = (0.3, 0.5, 0.2)^T$, respectively. The systems evolve in different attractors (**d**), that is, the trajectories of the systems are not in synchrony despite the coupling. However, each pair of projections (**a**), (**b**), (**c**) appears to be a distorted (nonlinear) image of one another [28]. Note that the difference between the trajectories (**e**) of both systems $e_x(t) = x_1(t) - x_2(t)$, $e_y(t) = y_1(t) - y_2(t)$ and $e_z(t) = z_1(t) - z_2(t)$ does not converge to zero as $t \to \infty$ nor is there a synchronous state that satisfies relationships $x_1 = x_2$, $y_1 = y_2$ and $z_1 = z_2$

In the synchronization of different chaotic systems, the differential equation associated with each system is completely different from one another. It is said that these coupled systems can be synchronized in a generalized sense if the trajectories of the coupled systems are connected by a differentially continuous mapping.[7] Clearly, this mapping defines the synchronization manifold, e.g., for Rössler and Lorenz coupled systems with trajectories $x = (x_1 x_2)$, $y = (y_1 y_2)$ and $z = (z_1 z_2)$, we have

$$\mathcal{M} = \left\{(x, y, z)|x_1 = \phi_x(x_2), y_1 = \phi_y(y_2), z_1 = \phi_z(z_2)\right\}.$$

Therefore, the definition of $\mathcal{M}$ depends on the existence of a mapping $\phi(\cdot, \cdot, \cdot) = \phi_x(\cdot), \phi_y(\cdot), \phi_z(\cdot)$. However, finding and defining $\phi(\cdot, \cdot, \cdot)$ is not trivial. This phenomenon was first observed by Afraimovich et al. [3] in 1986. Later it was called *generalized synchronization* (GS) in 1995 by Rulkov et al. [55] after studying the unidirectional coupling in the master–slave configuration of different chaotic systems. It is important to note that the master–slave configuration allows this problem to be treated in more general schemes of mutual coupling,[8] However, the mapping that defines the synchronization manifold may not exist. A particular case happens when the mapping is the identity, as a result of the coupled systems being identical, this is known as complete synchronization (CS). In general terms, the problem of GS is divided into two parts: (1) find explicitly the mapping that relates the trajectories of the slave system with those of the master system and (2) verify the stability of the synchronization manifold.

From the above, the existence of such transformation in the synchronization between different chaotic systems is essential. Invertibility and smoothness are problems inherent to the transformation since it must be a diffeomorphism. If it exists, it is guaranteed that the trajectories are confined in a synchronization manifold, as long as it is attractive. In the literature there are tools that allow detecting the existence of the GS phenomenon (auxiliary system approach [2, 28, 50], mutual false nearest neighbors [55], blending chaotic attractors [1], modified system approach [21], etc.). The stability analysis from these tools is based on studying the variational equation associated with the transversal manifold. However, the problems inherent to the analysis of the stability of the transversal manifold restrict these tools, e.g., the calculation of the maximum Lyapunov exponent [16, 45].

However, the works focused on finding the mapping are scarce and restrictive [26, 52, 66, 69], In addition, when this mapping is proposed, it is unknown if the synchronization manifold is indeed attractive [28]. Also, the tools mentioned in the

[7] There are many ways to define a mapping for this purpose [9]: continuous [2, 55], homeomorphism [3], differentiable continuous [23], diffeomorphism [28, 33, 36, 47]. Here, a differentially continuous mapping resulting from the use of tools based on differential algebra for multi-synchronization is of particular interest [10, 11, 36].

[8] There exist many ways to couple chaotic systems. The coupling used determines the complexity of the problem (as a result, the control law used to bring the systems to the synchronization manifold is more complex as well). The most common types of couplings are: complete replacement [44], partial replacement [12, 13], diffusive-coupling [3, 18, 54], state observers [38, 57], active control [4, 7], etc.

previous paragraph omit how to find this mapping, that is, they do not indicate how the trajectories of the master and slave systems are related. Instead, they only determine the existence of the mapping (in general, these methods transform the GS detection problem into a CS detection problem). Consequently, the form of the synchronization manifold is not obtained. This is not critical, but it certainly limits its practical utility.

1.6 Differential Algebra as a Solution

The tools based on differential algebra and especially the existence of the *differential primitive element* allow solving different problems associated with the construction of state observers and control laws for solving problems such as monitoring and local stabilization, trajectory tracking, systems tolerant to failures, simultaneous state and parameters estimation, etc. [14, 33, 35]. Taking into consideration that the synchronization problem is similar to the tracking problem, it is reasonable to think that this problem can be solved by employing a differential algebraic approach.

That is the case in this book, where the key to facing the synchronization problem is a differential primitive element of each coupled system (e.g., the output of each system). From this family of elements, we can obtain naturally the mapping that connects the coupled systems with the synchronized trajectories (the diffeomorphism that relates the trajectories of the slave system with those of the master). As a consequence, we can accurately describe the synchronization manifold, and therefore, the GS problem is partially solved. Preliminary work in this direction can be found in [34, 36].

1.7 Multi-synchronization

Motivated by phenomena observed in nature, especially in multiple oscillators [15, 53, 59], the problems associated with the synchronization of a pair of coupled systems have been extended to the study of increasingly complex scenarios. As a result, numerous applications of these problems are of great interest among the scientific community: “rendezvous”, training control, “flocking”, “schooling”, “attitude alignment”, sensor networks, distributed computing, consensus, complex networks in general, synchronization of chaos [8, 19, 31, 39, 53, 56] . In a current context, the synchronization of multiple coupled systems (*multi-synchronization*) can be understood in the following manner: In a network of systems called *multi-agent system* (MAS), formed by individuals (agents) that share total or partial information of their state with their neighboring systems, these can find a particular trajectory to follow according to the interactions described by an individual protocol (coupling) that depends on the information shared on the network (e.g., the output or state of each system). The way to model the interactions of the agents is given by a graph that indicates the topology of the interaction of the systems in the MAS.[9] Without loss of generality, we can talk about topology with or without a leader. The first is understood as the unidirectional configuration of a master and several slaves, in this way the trajectories of all slave systems (i.e., followers) follow a single master system (i.e., an acyclic and connected directed graph where there is a path of the root node or leader to all and each one of the nodes in the graph). In the literature, this problem is known as *consensus following the leader*. In the second, there is no master (there are no leader or followers) and there may be any type of connection between the agents in the network (e.g., graphs of small-world networks [67], graphs not directed in general, etc.); the trajectories of all systems (nodes) follow a common path that does not necessarily belong to any of the agents in the network. The object of study in the problems of multi-synchronization is focused on increasing the complexity of the interactions between the systems in the network. Works in this direction for nonlinear systems can be found in [5, 18, 25, 30, 42, 51, 58], all of them strongly motivated by discoveries in linear systems [31, 37, 40, 53].

The problems closest to reality are those where the systems that make up the MAS are different (i.e., networks of heterogeneous systems). In this area we can mention the following works [18, 25, 42, 43, 51, 64, 68].[10] Some of these results have been given in terms of practical stability of sets (attractors) that resemble in some way the analysis of the stability of the synchronization manifold. As far as we know, there are no results where the synchronization manifold is obtained explicitly. In this book, one of the objectives is to study the problem of networks of heterogeneous chaotic systems, that is, the problem of multi-synchronization in a generalized sense, which we call *generalized multi-synchronization* (GMS).

Now, the definition of the problem lies mainly in the existence of differentially continuous mappings that relate the trajectories of the master system with each of the slave systems, which restricts us from modeling the interaction between systems with a graph directed invariant in time (i.e., in the unidirectional scheme) not strongly connected (i.e., there is no sequence of axes from one node to any other node), commonly known in the literature as a *directed tree of coverage with a single leader as a root*. From the above, the fact that the graph is somewhat trivial does not imply that the GMS problem is as well trivial. This is quite clear since it is necessary to find the mappings and verify the stability of the synchronization manifold, which is now of a higher order than in the case of GS. For strictly different systems, the synchronization manifold is not trivial and may not exist. Furthermore, little is known about the stability of said variety and as a consequence, the synchronization of said systems [10, 11, 34, 42].

[9] Each node in the graph represents the dynamics associated with each of the systems in the network, for more detail on Graph Theory (see [17]).

[10] Also, according to the nature of the problem, it is possible to have intermittent connections between systems that are unavoidable [43, 64]. This possibility is beyond the scope of this book.

1.8 GMS and the Algebraic-Differential Approach

The GMS problem for systems in a master–slave configuration will be tackled in this book. As has been said, this problem can be reinterpreted as a consensus problem following the leader of MAS. Thus, it will be considered that a MAS is a network of chaotic, interconnected, heterogeneous systems of the same dimension, whose interactions are modeled by a fixed directed graph and not strongly connected as shown in Fig. 1.6.

From a differential-algebraic point of view and under the premise of the existence of a differential primitive element for each system, it is possible to give a solution to this problem in the following way: Through a family of transformations (diffeomorphic) ϕ_i obtained by the differential primitive element for each system, the MAS can be carried to a *multi-output generalized observability canonical form* (MOGOCF), see Fig. 1.7. The proper selection of the differential primitive element is key to solving this problem. This element is chosen as a linear combination of available states and control inputs (these elements allow the synchronization manifold to be explicitly obtained). In this transformed space is possible to design a dynamic protocol of consensus with the output of the leader, such that the trajectories of the followers eventually approach the trajectories of the leader, and by obtaining the differentially continuous mapping that defines the synchronization manifold (depending on the diffeomorphic transformations ϕ_i), a consensus to follow the leader of the MAS is reached when the said manifold is attractive. That is to say, all the nodes in the network approach a common path described by the leader as $t \to \infty$. As a result, a GMS state of the MAS is reached.

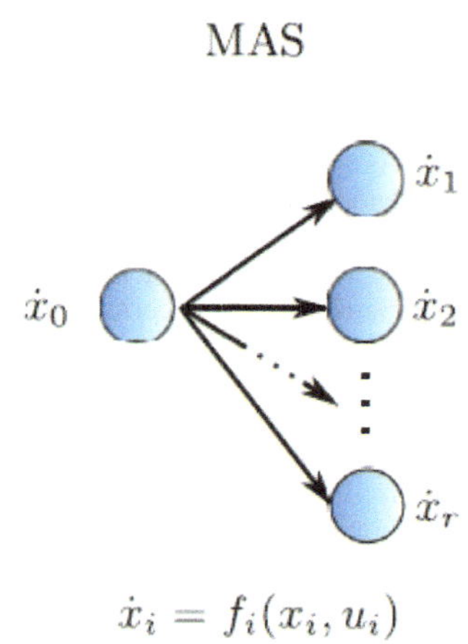

Fig. 1.6 MAS with a single leader, without interaction between followers. The graph associated is a coverage tree directed with a unique leader as a root. Each node represents the individual dynamics associated with each of the $r + 1$ interconnected systems

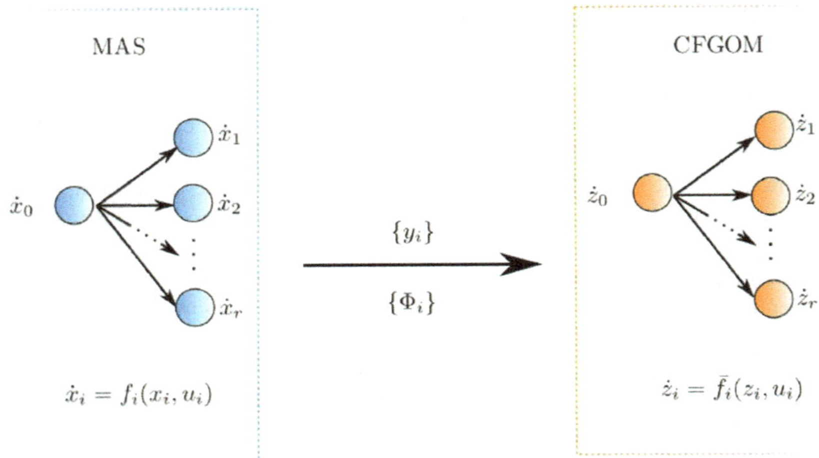

Fig. 1.7 MAS and GMS through a differential algebraic approach. $\{y_i\}$ is the family of differential primitive elements, and $\{\phi_i\}$ is the family of transformations

References

1. Abarbanel, H.D., Rulkov, N.F., Sushchik, M.M.: Blending chaotic attractors using the synchronization of chaos. Phys. Rev. E **52**(1), 214 (1995)
2. Abarbanel, H.D., Rulkov, N.F., Sushchik, M.M.: Generalized synchronization of chaos: the auxiliary system approach. Phys. Rev. E **53**(5), 4528 (1996)
3. Afraimovich, V., Verichev, N., Rabinovich, M.I.: Stochastic synchronization of oscillation in dissipative systems. Radiophys. Quantum Electron. **29**(9), 795–803 (1986)
4. Agiza, H., Yassen, M.: Synchronization of Rossler and Chen chaotic dynamical systems using active control. Phys. Lett. A **278**(4), 191–197 (2001)
5. Arcak, M.: Passivity as a design tool for group coordination. IEEE Trans. Autom. Control **52**(8), 1380–1390 (2007)
6. Ashwin, P., Buescu, J., Stewart, I.: Bubbling of attractors and synchronisation of chaotic oscillators. Phys. Lett. A **193**(2), 126–139 (1994)
7. Bai, E.-W., Lonngren, K.E.: Synchronization of two Lorenz systems using active control. Chaos, Solitons Fractals **8**(1), 51–58 (1997)
8. Barahona, M., Pecora, L.M.: Synchronization in small-world systems. Phys. Rev. Lett. **89**(5), 054101 (2002)
9. Boccaletti, S., Kurths, J., Osipov, G., Valladares, D., Zhou, C.: The synchronization of chaotic systems. Phys. Rep. **366**(1–2), 1–101 (2002)
10. Cruz-Ancona, C.D., Martínez-Guerra, R.: Fractional dynamical controllers for generalized multi-synchronization of commensurate fractional order Liouvillian chaotic systems. J. Franklin Inst. **354**(7), 3054–3096 (2017)
11. Cruz-Ancona, C.D., Martínez-Guerra, R., Pérez-Pinacho, C.A.: Generalized multi-synchronization: a leader-following consensus problem of multi-agent systems. Neurocomputing **233**, 52–60 (2017)
12. Cuomo, K.M., Oppenheim, A.V.: Circuit implementation of synchronized chaos with applications to communications. Phys. Rev. Lett. **71**(1), 65 (1993)
13. Cuomo, K.M., Oppenheim, A.V., Strogatz, S.H.: Synchronization of Lorenz-based chaotic circuits with applications to communications. IEEE Trans. Circ. Syst. II: Analog Digit. Signal Process. **40**(10), 626–633 (1993)

14. Martínez-Guerra, R., Cruz-Ancona, C.D.: Algorithms of Estimation for Nonlinear Systems: A Differential and Algebraic Viewpoint. Springer (2019)
15. Dörfler, F., Bullo, F.: Synchronization in complex networks of phase oscillators: a survey. Automatica **50**(6), 1539–1564 (2014)
16. Eckmann, J.-P., Ruelle, D.: Ergodic theory of chaos and strange attractors. In: The Theory of Chaotic Attractors, pp. 273–312. Springer (1985)
17. Godsil, C., Royle, G.F.: Algebraic Graph Theory, vol. 207. Springer Science & Business Media (2013)
18. Hale, J.K.: Diffusive coupling, dissipation, and synchronization. J. Dyn. Differ. Eq. **9**(1), 1–52 (1997)
19. Heagy, J., Carroll, T., Pecora, L.: Desynchronization by periodic orbits. Phys. Rev. E **52**(2), R1253 (1995)
20. Heagy, J.F., Pecora, L.M., Carroll, T.L.: Short wavelength bifurcations and size instabilities in coupled oscillator systems. Phys. Rev. Lett. **74**(21), 4185 (1995)
21. Hramov, A.E., Koronovskii, A.A.: Generalized synchronization: a modified system approach. Phys. Rev. E **71**(6), 067201 (2005)
22. Huang, L., Chen, Q., Lai, Y.-C., Pecora, L.M.: Generic behavior of master-stability functions in coupled nonlinear dynamical systems. Phys. Rev. E **80**(3), 036204 (2009)
23. Hunt, B.R., Ott, E., Yorke, J.A.: Differentiable generalized synchronization of chaos. Phys. Rev. E **55**(4), 4029 (1997)
24. Huygens, C.: Letter to de sluse. Letter no. 1333 of February 24, 1665. Oeuvres Complète de Christiaan Huygens. Correspondence 5(1665), 1664–1665
25. Isidori, A., Marconi, L., Casadei, G.: Robust output synchronization of a network of heterogeneous nonlinear agents via nonlinear regulation theory. IEEE Trans. Autom. Control **59**(10), 2680–2691 (2014)
26. Juan, M., Xingyuan, W.: Generalized synchronization via nonlinear control. Chaos Interdiscip. J. Nonlinear Sci. **18**(2), 023108 (2008)
27. Khalil, H.K., Grizzle, J.W.: Nonlinear Systems, vol. 3. Prentice Hall, Upper Saddle River, NJ (2002)
28. Kocarev, L., Parlitz, U.: Generalized synchronization, predictability, and equivalence of unidirectionally coupled dynamical systems. Phys. Rev. Lett. **76**(11), 1816 (1996)
29. Layek, G.: An Introduction to Dynamical Systems and Chaos. Springer (2015)
30. Lin, Z., Francis, B., Maggiore, M.: State agreement for continuous-time coupled nonlinear systems. SIAM J. Control Optim. **46**(1), 288–307 (2007)
31. Lynch, N.A.: Distributed Algorithms. Elsevier (1996)
32. Martínez-Guerra, R., García, J.J.M., Prieto, S.M.D.: Secure communications via synchronization of Liouvillian chaotic systems. J. Franklin Inst. **353**(17), 4384–4399 (2016)
33. Martínez-Guerra, R., Gómez-Cortés, G., Pérez-Pinacho, C.: Synchronization of integral and fractional order chaotic systems. In: A Differential Algebraic and Differential Geometric Approach. Springer (2015)
34. Martínez-Guerra, R., Mata-Machuca, J.L.: Fractional generalized synchronization in a class of nonlinear fractional order systems. Nonlinear Dyn. **77**(4), 1237–1244 (2014)
35. Martinez-Guerra, R., Mata-Machuca, J.L.: Fault Detection and Diagnosis in Nonlinear Systems. Springer (2016)
36. Martínez-Guerra, R., Pérez-Pinacho, C.A., Gómez-Cortés, G.C.: Generalized synchronization via the differential primitive element. In: Synchronization of Integral and Fractional Order Chaotic Systems, pp. 163–174. Springer (2015)
37. Mesbahi, M., Egerstedt, M.: Graph Theoretic Methods in Multiagent Networks, vol 33. Princeton University Press (2010)
38. Nijmeijer, H., Mareels, I.M.: An observer looks at synchronization. IEEE Trans. Circ. Syst. I: Fundam. Theory Appl. **44**(10), 882–890 (1997)
39. Nishikawa, T., Motter, A.E.: Symmetric states requiring system asymmetry. Phys. Rev. Lett. **117**(11), 114101 (2016)

40. Olfati-Saber, R., Fax, J.A., Murray, R.M.: Consensus and cooperation in networked multi-agent systems. Proc. IEEE **95**(1), 215–233 (2007)
41. Ott, E., Sommerer, J.C.: Blowout bifurcations: the occurrence of riddled basins and on-off intermittency. Phys. Lett. A **188**(1), 39–47 (1994)
42. Panteley, E., Loria, A.: On practical synchronisation and collective behaviour of networked heterogeneous oscillators. IFAC-PapersOnLine **48**(18), 25–30 (2015)
43. Panteley, E., Loría, A.: Synchronization and dynamic consensus of heterogeneous networked systems. IEEE Trans. Autom. Control **62**(8), 3758–3773 (2017)
44. Pecora, L.M., Carroll, T.L.: Synchronization in chaotic systems. Physical Rev. Lett. **64**(8), 821 (1990)
45. Pecora, L.M., Carroll, T.L.: Driving systems with chaotic signals. Phys. Rev. A **44**(4), 2374 (1991)
46. Pecora, L.M., Carroll, T.L.: Master stability functions for synchronized coupled systems. Phys. Rev. Lett. **80**(10), 2109 (1998)
47. Pecora, L.M., Carroll, T.L., Heagy, J.F.: Statistics for mathematical properties of maps between time series embeddings. Phys. Rev. E **52**(4), 3420 (1995)
48. Pecora, L.M., Carroll, T.L., Johnson, G.A., Mar, D.J., Heagy, J.F.: Fundamentals of synchronization in chaotic systems, concepts, and applications. Chaos Interdiscip. J. Nonlinear Sci. **7**(4), 520–543 (1997)
49. Pikovsky, A.S., Grassberger, P.: Symmetry breaking bifurcation for coupled chaotic attractors. J. Phys. A: Math. General **24**(19), 4587 (1991)
50. Pyragas, K.: Weak and strong synchronization of chaos. Phys. Rev. E **54**(5), R4508 (1996)
51. Qu, Z., Chunyu, J., Wang, J.: Nonlinear cooperative control for consensus of nonlinear and heterogeneous systems. In: 2007 46th IEEE Conference on Decision and Control, pp. 2301–2308. IEEE (2007)
52. Ramírez, J.B., Galarza, K.C., Femat, R.: Generalized synchronization of strictly different systems: partial-state synchrony. Chaos, Solitons Fractals **45**(3), 193–204 (2012)
53. Ren, W., Beard, R.W., Atkins, E.M.: Information consensus in multivehicle cooperative control. IEEE Control Syst. Mag. **27**(2), 71–82 (2007)
54. Rul'Kov, N., Volkoskii, A., Rodriguez-Lozano, A., Del Rio, E., Velarde, M.: Mutual synchronization of chaotic self-oscillators with dissipative coupling. Int. J. Bifurc. Chaos **2**(03), 669–676 (1992)
55. Rulkov, N.F., Sushchik, M.M., Tsimring, L.S., Abarbanel, H.D.: Generalized synchronization of chaos in directionally coupled chaotic systems. Phys. Rev. E **51**(2), 980 (1995)
56. Russo, G., Di Bernardo, M.: Contraction theory and master stability function: linking two approaches to study synchronization of complex networks. IEEE Trans. Circ. Syst. II: Express Briefs **56**(2), 177–181 (2009)
57. So, P., Ott, E., Dayawansa, W.: Observing chaos: deducing and tracking the state of a chaotic system from limited observation. Phys. Rev. E **49**(4), 2650 (1994)
58. Stan, G.-B., Sepulchre, R.: Analysis of interconnected oscillators by dissipativity theory. IEEE Trans. Autom. Control **52**(2), 256–270 (2007)
59. Strogatz, S.H.: Exploring complex networks. Nature **410**(6825), 268 (2001)
60. Strogatz, S.H.: Sync: How Order Emerges from Chaos in the Universe, Nature, and Daily Life. Hachette, UK (2012)
61. Strogatz, S.H.: Nonlinear Dynamics and Chaos with Student Solutions Manual: With Applications to Physics, Biology, Chemistry, and Engineering. CRC Press (2018)
62. Sun, J., Bollt, E.M., Nishikawa, T.: Master stability functions for coupled nearly identical dynamical systems. EPL (Europhys. Lett.) **85**(6), 60011 (2009)
63. Tang, Y., Mees, A., Chua, L.: Synchronization and chaos. IEEE Trans. Circ. Syst. **30**(9), 620–626 (1983)
64. Wang, L., Chen, M.Z., Wang, Q.-G.: Bounded synchronization of a heterogeneous complex switched network. Automatica **56**, 19–24 (2015)
65. Wang, X.-Y., Wang, M.-J.: A chaotic secure communication scheme based on observer. Commun. Nonlinear Sci. Numer. Simul. **14**(4), 1502–1508 (2009)

66. Wang, Y.-W., Guan, Z.-H.: Generalized synchronization of continuous chaotic system. Chaos, Solitons Fractals **27**(1), 97–101 (2006)
67. Watts, D.J., Strogatz, S.H.: Collective dynamics of small-world networks. Nature 393(6684), 440 (1998)
68. Xu, D., Wang, X., Hong, Y., Jiang, Z.-P.: Global robust distributed output consensus of multi-agent nonlinear systems: an internal model approach. Syst. Control Lett. **87**, 64–69 (2016)
69. Yang, S., Duan, C.: Generalized synchronization in chaotic systems. Chaos, Solitons Fractals **9**(10), 1703–1707 (1998)
70. Zhang, Y., Nishikawa, T., Motter, A.E.: Asymmetry-induced synchronization in oscillator networks. Phys. Rev. E **95**(6), 062215 (2017)

Chapter 2
Synchronization of Non-identical Systems

In the previous chapter, it was briefly discussed the challenges to achieve and explain the synchronization of strictly different chaotic systems. As a solution, was introduced the Generalized Synchronization (GS) concept [17]. This fundamental phenomena is currently widely studied, having both theoretical and applied significance [2, 5, 8, 11, 15, 18–20].

GS occurs when the trajectories of master and slave systems are equal through a mapping. From the differential algebra point of view, GS appears if there exists a *differential primitive element* such that generates a mapping H_{ms} from the trajectories $x_m(t)$ of the attractor in the master algebraic manifold $\mathcal{M}$ to the trajectories $x_s(t)$ in the slave space R^{n_s}, i.e., $H_{ms}(x_s(t)) = x_m(t)$. For identical systems, the functional mapping corresponds to the identity [12].

This book aims to introduce a method for GS in nonlinear systems, where it is sufficient to know the output[1] of the system to generate this transformation which is represented using a differential transcendence basis, that is to say, there exists an element $\bar{y}$ and let $n \neq 0$ be the minimum integer such that $\bar{y}^{(n)}$ is analytically dependent on $\bar{y}, \bar{y}^{(1)}, \ldots, \bar{y}^{(n-1)}$ such that

$$\bar{H}(\bar{y}, \bar{y}^{(1)}, \ldots, \bar{y}^{(n-1)}, \bar{y}^{(n)}, u, u^{(1)}, \ldots, u^{(\gamma)}) = 0 \tag{2.1}$$

The main idea is to find a dynamical control signal such that it is possible to synchronize the system in a new coordinates representation. In other words, the original system is carried out to a triangular form through an adequate choice of the

[1] In general, the state vector of a dynamic system is partially known, due to economic or technological restrictions. Therefore, here we consider the synchronization of systems with certain states available from measurements.

R. Martínez-Guerra and J. P. Flores-Flores, *An Approach to Multi-agent Systems as a Generalized Multi-synchronization Problem*, Understanding Complex Systems,
https://doi.org/10.1007/978-3-031-22669-4_2

differential primitive element, which is normally chosen as a linear combination of the known states and the inputs of the system and whose coefficients belong to a differential field.

2.1 GS: Problem Statement

Let us consider two strictly different chaotic systems. We aim to synchronize both systems through a unidirectional coupling signal, this is, we consider a master–slave configuration. In particular, consider the following diffusively coupled Rössler and Lorenz systems, whose coupling g is at the first component:

$$\begin{aligned}\dot{x}_1 &= -(y_1 + z_1),\\ \dot{y}_1 &= x_1 + 0.2y_1,\\ \dot{z}_1 &= 0.2 + z_1(x_1 - \mu)\end{aligned}$$

and

$$\begin{aligned}\dot{x}_2 &= \sigma(y_2 - x_2) - g(x_2 - x_1),\\ \dot{y}_2 &= -x_2z_2 + rx_2 - y_2,\\ \dot{z}_2 &= x_2y_2 - bz_2\end{aligned}$$

with parameters $\sigma, r, b, \mu > 0$, and coupling strength $g \geq 0$. To reach a synchronous state, there exist several possibilities. In the following, we will describe some methodologies, so that we can contrast them with the differential-algebraic approach.

2.2 Variational Methods for GS

In the early stages of its development, generalized synchronization was based on the properties of the variational equation related to the synchronization error. Hence, variational methods were developed to determine whether a pair of coupled systems possesses this property. In what follows, some of these methods are described to grasp the essence of the problem.

2.2.1 False Nearest Neighbors

This method was developed in [7] to determine the minimum embedding dimension necessary to reconstruct the state space of a dynamical system from observed data with time delay. The method was later introduced by Rulkov et.al. to chaos synchronization theory [17] to detect the presence of the continuous transformation $\Phi(\cdot)$ and thereby distinguish between synchronized and unsynchronized behaviors of coupled chaotic systems.

The false nearest neighbors method is based on the concept of local neighborliness. This is, given a master and a slave system, when the trajectories in the phase space of each system are related, i.e., $y(t) = \Phi(x(t))$, two close states in the phase space of the response system correspond to two close states in the space of the driving system.

Thus, the method aims to find a connection between the systems. The kind of relation found with this method is geometrical, and it can be summarized as follows. Considering an arbitrary point $x_n = x(t_n)$ in the phase space of the master system, the nearest phase space neighbor of this point has a time index n_{NND}. At the same time, in the phase space of the slave system, the point y_n is associated with a close neighbor point $y_{n_{NND}}$. Thus, the distances between the two nearest neighbors in the phase spaces of the master and slave systems can be expressed as

$$y_n - y_{n_{NND}} = \Phi(x_n) - \Phi(x_{n_{NND}}) \approx \frac{\partial \Phi(x_n)}{\partial x_n}(x_n - x_{n_{NND}})$$

Similarly, we go to time index n and observe the slave vector y_n and locate its nearest neighbor y_n which comes at time index n_{NNR}. Therefore, in a similar way

$$y_n - y_{n_{NNR}} = \Phi(x_n) - \Phi(x_{n_{NNR}}) \approx \frac{\partial \Phi(x_n)}{\partial x_n}(x_n - x_{n_{NNR}})$$

Thus, the ratio

$$\frac{|y_n - y_{n_{NND}}|\,|y_n - y_{n_{NNR}}|}{|x_n - x_{n_{NND}}|\,|x_n - x_{n_{NNR}}|}$$

known as the mutual false nearest neighbors parameter determines the existence of a synchronous state of the type $y(t) = \Phi(x(t))$. The criterion is, that if this parameter is of order equal to one, then the systems are synchronized. Otherwise, if the synchronization relation $y(t) = \Phi(x(t))$ does not hold, the parameter should on average be of an order given by the quotient between the size of attractor squared and the distance between nearest neighbors squared.

2.2.2 *The Auxiliary System Approach*

Proposed by Abarbanel et al. (see [1]), this method is used to detect and characterize the forced generalized synchronization of chaotic systems. At the time, given the limited computational capabilities (1990s), it emerged as an alternative to the contemporaneous complex computational methods, as an option to implement in experiments [17].

As we know, generalized synchronization refers to the existence of a continuous transformation $\Phi(\cdot)$ between the synchronized trajectories of drive and response systems. Originally, the stability of $y(t) = \Phi(x(t))$ was studied to demonstrate the existence of a synchronous state, where $x(t) \in \mathscr{X}$ and $y(t) \in \mathscr{Y}$ are the state of master and slave systems, respectively. However, in the vast majority of cases, this analysis is extremely complicated since requires the use of numerical methods. Consider the following master and slave systems

$$\dot{x}(t) = F(x(t)),$$
$$\dot{y}(t) = G(y(t), g, x(t))$$

with coupling strength g and subject to the following conditions

(1) $\Phi(\cdot)$ has no explicit time dependence.
(2) The transformation that takes points in the $\mathscr{X}$ space to points (not continuous domains) in the $\mathscr{Y}$ space does not require preserving the number of points operated upon.
(3) In each branch, the transformation is locally continuous.

The auxiliary system approach method states that considering an auxiliary system

$$\dot{z}(t) = G(z(t), g, x(t))$$

which is identical to the slave system and is driven by the same signal $x(t)$, the stability analysis of the synchronization manifold in the space $\mathscr{X} \oplus \mathscr{Y}$, which in general may have a very complicated shape $y(t) = \Phi(x(t))$, can be replaced by the stability analysis of the quite simple auxiliary manifold $z(t) = y(t)$ in $\mathscr{Z} \oplus \mathscr{Y}$ space.

When one can prove the local stability of the auxiliary manifold, the conclusion that follows is that in the combined phase space of the master and slave systems there exists an attractor that is the image of generalized synchronized chaotic oscillations. This approach has been deeply explored by some other authors, who have managed to establish stability criteria based on Lyapunov exponents and matrix measure [3, 21].

2.2.3 Modify System Approach

Given master and slave systems, a synchronous state might depend on the coupling strength parameter. This problem can be also analyzed using a bifurcation theory approach. Modify system approach considers two unidirectionally coupled chaotic oscillators, that is

$$\dot{x}_d = F(x_d),$$
$$\dot{y}_r = G(x_r) + gH(x_d - x_r)$$

where x_d and y_r are the states of master and slave systems, respectively, g is the coupling strength parameter and H is called the coupling matrix. Along with these, there is an additional system called the modified system, described by

$$\dot{x}_m = G'(x_m, g) + gHx_d$$

This system is used to describe the GS arising [4], this is, two cooperative processes take place simultaneously. The first of them is the growth of the dissipation in the modified system and the second one is the increasing of the amplitude of the external signal, both processes depend on g. In the first process, g is a dissipation parameter, such that simplifies the dynamics of the modified system. The last goes from chaotic oscillations to periodic ones. Also, if the dissipation is large enough, it might show a stationary state (see [16]). On the other hand, the external signal tends to impose the dynamics of the master chaotic oscillator on the modified system. This effect is suppressed by dissipation. Thus, GS might take place only if the proper chaotic dynamics of the modified system are suppressed by the dissipation, this is, modified system dynamics is completely determined by $x_m = \Phi(x_d)$ and therefore $x_r = \Phi(x_d)$. In conclusion, GS arising in the response system is possible for such values of g such that the modified system demonstrates the periodic oscillations or the stationary state.

2.3 The Differential Primitive Element

An alternative for studying the GS problem is the differential-algebraic approach, whose key is the usage of the so-called differential primitive element. However, before introducing this methodology, let us consider the following definitions[2]:

Definition 2.1 A differential field K is a set with the properties of a field that is additionally endowed with a single derivation that obeys the usual rules, i.e.,

$$\frac{d}{dt}(a+b) = \frac{da}{dt} + \frac{db}{dt}, \quad \forall a, b \in \mathrm{K}$$
$$\frac{d}{dt}(a \cdot b) = \frac{da}{dt}b + a\frac{db}{dt}, \quad \forall a, b \in \mathrm{K}$$

Example 2.1 The trivial sets $\mathbb{Q}$, $\mathbb{R}$ and $\mathbb{C}$ are differential fields.

Example 2.2 Let $R[x]$ be the ring of polynomials with single variable x over field $\mathbb{R}$ and single derivation $\partial : R[x] \to R[x]$, then $R[x]$ is a differential field.

[2] The differential-algebraic approach has been successfully used to tackle relevant topics such fail detection, nonlinear observers design, etc. Further details can be found in [10, 13].

Definition 2.2 A differential field extension L/K is given by two differential fields K and L such that: (a) $\mathrm{K} \subset \mathrm{L}$, and (b) the derivation of K is the derivation of L restricted to K.

Example 2.3 Let $\mathrm{L} = \mathbb{R} < e^t >$, $\mathrm{K} = \mathbb{R}$. Then $\mathbb{R} < e^t > /\mathbb{R}$ is a differential field extension.

Definition 2.3 An element $a \in \mathrm{L}$ is said to be differentially algebraic over K if and only if it satisfies a differential equation $P(a, \dot{a}, \ldots, a^{(\gamma)}) = 0$, with P a polynomial over K in a and its time derivatives, $\gamma \in \mathbb{Z}^+$. On the other hand, if a does not satisfy a differential equation we say that a is differentially transcendental over K.

Example 2.4 The element $a = e^t \in \mathrm{L}$ satisfies the differential equation $\dot{x} - x = 0$. Hence, $a = e^t$ is differentially algebraic over K.

An element $\delta \in \mathrm{L}$ such that $\mathrm{L} = \mathrm{K} < \delta >$ is called differential primitive element of the differential field extension L/K.

Definition 2.4 We define a dynamics as a finitely generated differential algebraic extension $\mathrm{L}/\mathrm{K} < u >$ of the differential field $\mathrm{K} < u >$, with $\mathrm{K} < u >$ the differential field generated by K and a finite set of differential quantities $u = (u_1, u_2, \ldots, u_\mu)$, $\mu \in \mathbb{Z}^+$.

Example 2.5 Consider the following differential equation:

$$\ddot{y} + \dot{y} + \sin y + u = 0$$

In this case y is algebraic over $\mathbb{R} < u >$, and it can be seen as a dynamics of the form $\mathbb{R} < u, y > /\mathbb{R} < u >$.

Definition 2.5 A system is Picard–Vessiot (PV) if and only if the $\mathrm{K} < u >$-vector space generated by the derivatives of the set $\{y^{(\mu)}, \mu \geq 0\}$ has finite dimension.

Then, the system (2.1) can be solved locally as

$$\bar{y}^{(n)} = -\mathcal{L}\left(\bar{y}, \ldots, \bar{y}^{(n-1)}, u, u^{(1)}, \ldots, u^{(\gamma-1)}\right) + u^{(\gamma)}$$

and recalling $\xi_i = \bar{y}^{(i-1)}$, $1 \leq i \leq n$. Then a local form is obtained, which can be seen as a GOCF (*generalized observability canonical form*),

$$\begin{aligned}
\dot{\xi}_1 &= \xi_2 \\
\dot{\xi}_2 &= \xi_3 \\
&\vdots \\
\dot{\xi}_{n-1} &= \xi_n \\
\dot{\xi}_n &= -\mathcal{L}\left(\xi_1, \ldots, \xi_n, u, u^{(1)}, \ldots, u^{(\gamma-1)}\right) + u^{(\gamma)} \\
\bar{y} &= \xi_1
\end{aligned} \tag{2.2}$$

Let us now consider the following class of nonlinear systems:

$$\begin{aligned} \dot{x} &= F(x, u) \\ y &= Cx \end{aligned} \tag{2.3}$$

where $x \in \mathbb{R}^n$, is the state vector, $F(\cdot)$ is a nonlinear vectorial function, u is the input, y is the output and C is a matrix of proper dimensions.

Lemma 2.1 *A nonlinear system* (2.3) *is transformable to a GOCF if and only if it is PV.*

Proof Let the set $\left\{\varepsilon, \varepsilon^{(1)}, \ldots, \varepsilon^{(n-1)}\right\}$ be a finite differential transcendence basis with $\varepsilon^{(i-1)} = y^{(i-1)}, 1 \leq i \leq n$ where $n \leq 0$ is the minimum integer such that $y^{(n)}$ is dependent on $y, y^{(1)}, \ldots, y^{(n-1)}, u, \ldots$

Redefining $\xi_i = \varepsilon^{(i-1)}$, $1 \leq i \leq n$, yields to

$$\begin{aligned} \dot{\xi}_j &= \xi_{j+1}, \quad 1 \leq j \leq n-1 \\ \dot{\xi}_n &= -\mathscr{L}\left(\xi_1, \ldots, \xi_n, u, \dot{u}, \ldots, u^{(\gamma-1)}\right) + u^{(\gamma)} \\ y &= \xi_1 \end{aligned} \tag{2.4}$$

Such that the lemma is proven. □

Now, we can discuss the GS of nonlinear systems which are completely triangularizable (GOCF). For this class of systems, the GS problem is solvable in the sense of [9, 17], using a dynamic feedback controller that stabilizes the synchronization error dynamics. The GS problem is stated as follows:

Suppose two nonlinear systems in a master–slave configuration, where the master system is given by

$$\begin{aligned} \dot{x}_m &= F_m\left(x_m, u_m\right) \\ y_m &= h_m\left(x_m\right) \end{aligned} \tag{2.5}$$

and the slave by

$$\begin{aligned} \dot{x}_s &= F_s\left(x_s, u_m\left(x_s, y_m\right)\right) \\ y_s &= h_s\left(x_s\right) \end{aligned} \tag{2.6}$$

where $x_s = (x_{1s}, \ldots, x_{n_s}) \in \mathbb{R}^{n_s}$, $x_m = (x_{1m}, \ldots, x_{n_m}) \in \mathbb{R}^{n_m}$, $h_s : \mathbb{R}^{n_s} \to \mathbb{R}$, $h_m : \mathbb{R}^{n_m} \to \mathbb{R}$, $u_m = (u_{1m}, \ldots, u_{\bar{m}_n}) \in \mathbb{R}^{\bar{m}_n}$, $u_s : \mathbb{R}^{n_s} \times \mathbb{R} \to \mathbb{R}$, y_m, $y_s \in \mathbb{R}$. Also, F_s, F_m, h_s and h_m are assumed to be polynomial in their arguments.

The last is a more general case, because the systems (2.5) and (2.6) are not necessarily affine nonlinear systems. Indeed, the dynamics of the slave system do not need to be expressed as a linear part and a nonlinear part as in [6], where the nonlinear vector function is restricted to satisfy a Lipschitz condition.

Definition 2.6 (Generalized Synchronization.) The slave and master systems are said to be in a state of generalized synchronization (GS) if there exists a differential primitive element which generates a mapping $H_{ms} : \mathbb{R}^{n_s} \rightarrow \mathbb{R}^{n_m}$ with $H_{ms} = \phi_m^{-1} \circ \phi_s$ as well as there exists an algebraic manifold $\mathscr{M} = \{(x_s, x_m) | x_m = H_{ms}(x_s)\}$ and a compact set $\mathscr{B} \subset \mathbb{R}^{n_m} \times \mathbb{R}^{n_s}$ with $\mathscr{M} \subset \mathscr{B}$ such that their trajectories, with initial condition in $\mathscr{B}$, approach to $\mathscr{M}$ as $t \rightarrow \infty$.

Definition 2.6 leads us to the following criterion:

$$\lim_{t \rightarrow \infty} ||H_{ms}(x_s) - x_m|| = 0 \tag{2.7}$$

Remark 2.1 Note that identical or complete synchronization (CS) is a particular case of GS that arises when the transformation $H_{ms} = I$.

Remark 2.2 The differential primitive element can be defined as

$$y = \sum_i \alpha_i x_i + \sum_j \beta_j u_j, \quad \alpha_i, \beta_j \in \mathrm{K} < u > \tag{2.8}$$

where $\mathrm{K} < u >$ is a differential field generated by K and u and their differential quantities.

Proposition 2.1 *Let systems* (2.5) *and* (2.6) *be transformable to a GOCF. Let us define* $z_m = (z_{m_1} \ldots, z_{m_n})^T$ *and* $z_s = (z_{s_1}, \ldots, z_{s_n})^T$ *as the trajectories of master and slave systems in the coordinate transformation, respectively, with* $z_{m_i} = y_m^{(i-1)}$ *and* $z_{s_i} = y_s^{(i-1)}$*, for* $1 \leq i \leq n$*. Moreover, assume that* Φ_m^{-1} *is a continuously differentiable (with uniformly bounded derivative). Then*

$$\lim_{t \rightarrow \infty} ||z_m - z_s|| = 0 \tag{2.9}$$

implies that

$$\lim_{t \rightarrow \infty} ||H_{ms}(x_s) - x_m|| = 0 \tag{2.10}$$

where y_m *and* y_s *are the differential primitive elements for the master and slave system, respectively.*[3]

Proof Without loss of generality we can choose $u_m = 0 \in \mathbb{R}^{\bar{m}_n}$. Then the differential primitive element for the master is taken as

$$y_m = \sum_i \alpha_{m_i} x_{m_i}, \quad \alpha_{m_i} \in \mathbb{R} < u_m > \tag{2.11}$$

[3] This proposition implies that CS and GS are achieved simultaneously, the first in the coordinate transformation and the last is obtained in the original coordinates.

and for the slave system:

$$y_s = \sum_i \alpha_{s_i} x_{s_i} + \sum_j \beta_{s_j} u_{s_j}, \quad \alpha_{s_i}, \beta_{s_i} \in \mathbb{R} < u_s > \tag{2.12}$$

which lead us to

$$\begin{aligned} \dot{z}_{m_j} &= z_{m_{j+1}}, \quad 1 \leq j \leq n-1 \\ \dot{z}_{m_n} &= -\mathscr{L}_m(z_{m_1}, \ldots, z_{m_n}) \end{aligned} \tag{2.13}$$

and

$$\begin{aligned} \dot{z}_{s_j} &= z_{s_{j+1}}, \quad 1 \leq j \leq n-1 \\ \dot{z}_{s_n} &= -\mathscr{L}_s\left(z_{s_1}, \ldots, z_{s_n}, u_s, \dot{u}_s, \ldots, u_s^{(\gamma-1)}\right) + u_s^{\gamma} \end{aligned} \tag{2.14}$$

Then, let us define the control signals as $u_1 = u_s$, $u_2 = \dot{u}_s$, ..., $u_\gamma = u^{(\gamma)}$, such that we can express the following dynamical system

$$\begin{aligned} \dot{u}_j &= u_{j+1}, \quad 1 \leq j \leq \gamma - 1 \\ \dot{u}_\gamma &= -\mathscr{L}_m(z_{m_1}, \ldots, z_{m_n}) + \mathscr{L}_s(z_{s_1}, \ldots, z_{s_n}, u_1, \ldots, u_\gamma) + \mathrm{K}(z_m - z_s) \end{aligned} \tag{2.15}$$

where $z_m = (z_{m_1}, \ldots, z_{m_n})^T$, $z_s = (z_{s_1}, \ldots, z_{s_n})^T$ and $\mathrm{K} = (k_1, \ldots, k_n)$. Therefore, the closed-loop dynamics of the synchronization error $e_z = z_m - z_s$ is given by the augmented system

$$\begin{aligned} \dot{e}_{z_j} &= e_{z_{j+1}}, \quad 1 \leq j \leq n-1 \\ \dot{e}_{z_n} &= -\mathscr{L}_m\left(z_{m_1}, \ldots, z_{m_n}\right) + \mathscr{L}_s\left(z_{s_1}, \ldots, z_{s_n}, u_1, \ldots, u_\gamma\right) - \dot{u}_\gamma \\ \dot{u}_j &= u_{j+1}, \quad 1 \leq j \leq \gamma - 1 \\ \dot{u}_\gamma &= -\mathscr{L}_m\left(z_{m_1}, \ldots, z_{m_n}\right) + \mathscr{L}_s\left(z_{s_1}, \ldots, z_{s_n}, u_1, \ldots, u_\gamma\right) + \mathrm{K}e_z \end{aligned} \tag{2.16}$$

Finally, we have that $\dot{e}_z = Ae_z$ with

$$A = \begin{pmatrix} 0 & 1 & 0 & 0 & \ldots & 0 \\ 0 & 0 & 1 & 0 & \ldots & 0 \\ \vdots & \vdots & \vdots & \ddots & \ldots & 0 \\ 0 & 0 & 0 & 0 & 1 & 0 \\ 0 & 0 & 0 & 0 & 0 & 1 \\ -k_1 & -k_2 & -k_3 & -k_4 & \ldots & -k_n \end{pmatrix} \tag{2.17}$$

where the control gains $(k_1, \ldots, k_n)$ are chosen such that the spectrum of $A \in \mathbb{R}^{n \times n}$ has only negative real parts. Thus $z_s \to z_m$ as $t \to \infty$. □

Corollary 2.1 *All system is in a state of GS if and only if it is PV.*

Proof The proof is trivial, and it has been omitted. □

2.4 Numerical Example

Consider the following Lorenz chaotic system as a master system

$$\begin{aligned}
\dot{x}_{1_L} &= a_L(x_{2_L} - x_{1_L}) \\
\dot{x}_{2_L} &= b_L x_{1_L} - x_{2_L} - x_{1_L} x_{3_L} \\
\dot{x}_{3_L} &= -c_L x_{3_L} + x_{1_L} x_{2_L}
\end{aligned} \tag{2.18}$$

where a_L, b_L and c_L are chosen such that (2.18) is chaotic. Let the differential primitive element be the output of system (2.18), that is, $y_L = x_{1_L}$. Thus, we propose the next coordinate transformation

$$\begin{bmatrix} z_{1_L} \\ z_{2_L} \\ z_{3_L} \end{bmatrix} = \begin{bmatrix} y_L \\ \dot{y}_L \\ \ddot{y}_L \end{bmatrix} = \begin{bmatrix} x_{1_L} \\ a_L(x_{2_L} - x_{1_L}) \\ a_L(b_L x_{1_L} - x_{2_L} - x_{3_L} x_{1_L} - a_L(x_{2_L} - x_{1_L})) \end{bmatrix} = \phi_L(x_L) \tag{2.19}$$

In the transformed coordinates, the Lorenz system (2.18) is rewritten as,

$$\begin{aligned}
\dot{z}_{1_L} &= z_{2_L} \\
\dot{z}_{2_L} &= z_{3_L} \\
\dot{z}_{3_L} &= \Psi_L(x_L)
\end{aligned} \tag{2.20}$$

where $\Psi_L(x_L) = \left(a_L^2 + a_L b_L - a_L x_{3_L}\right)\dot{x}_{1_L} - \left(a_L^2 + a_L\right)\dot{x}_{2_L} - a_L x_{1_L}\dot{x}_{3_L}$. Suppose the Chua chaotic system [14] as the controlled slave system,

$$\begin{aligned}
\dot{x}_{1_C} &= a_C(x_{2_C} - x_{1_C} - \nu_x) \\
\dot{x}_{2_C} &= x_{1_C} - x_{2_C} + x_{3_C} \\
\dot{x}_{3_C} &= -b_C x_{2_C}
\end{aligned} \tag{2.21}$$

where $\nu_x = m_0 x_{1_C} + \frac{1}{2}(m_1 - m_0)(|x_{1_C} + 1| - |x_{1_C} - 1|) \in \mathbb{R}$ and the parameters a_C, b_C, m_0, m_1 are chosen such that (2.21) is chaotic. For slave system, we assume the differential primitive element as the output of system (2.21) plus the control input, i.e., $y_C = x_{3_C} + u_1$. Using Proposition 2.1, the controlled coordinate transformation system (2.21) is rewritten as

$$\begin{bmatrix} z_{1_C} \\ z_{2_C} \\ z_{3_C} \end{bmatrix} = \begin{bmatrix} y_C \\ \dot{y}_C \\ \ddot{y}_C \end{bmatrix} = \begin{bmatrix} x_{3_C} + u_1 \\ -b_C x_{2_C} + u_2 \\ -b_C(x_{1_C} - x_{2_C} + x_{3_C}) + u_3 \end{bmatrix} = \phi_C(x_C) \tag{2.22}$$

where $u_1 = u$, $u_2 = \dot{u}$ and $u_3 = \ddot{u}$ are control signals which need to be designed to achieve synchronization between trajectories of coordinate transformation systems (2.19) and (2.22). The augmented controlled system in the transformed coordinates is represented as

$$\begin{aligned} \dot{z}_{1_C} &= z_{2_C} \\ \dot{z}_{2_C} &= z_{3_C} \\ \dot{z}_{3_C} &= \Psi x_C + \bar{u} \\ \dot{u}_1 &= u_2 \\ \dot{u}_2 &= u_3 \\ \dot{u}_3 &= \bar{u} \end{aligned} \tag{2.23}$$

where $\Psi_C(x_C) = -b(\dot{x}_{1_C} - \dot{x}_{2_C} + \dot{x}_{3_C})$. Then, the control objective is to find $\bar{u}$ such that the trajectories of slave system (2.23) follows the trajectories of master system (2.20). In other words, we need to find $\bar{u}$ of system (2.23), such that $(z_{1_C}, z_{2_C}, z_{3_C}) \to (z_{1_L}, z_{2_L}, z_{3_L})$ as long as $t \to \infty$.

Defining the synchronization error in the transformed coordinates as $e_z = z_L - z_C$, the error dynamics is given by

$$\begin{aligned} \dot{e}_1 &= e_2 \\ \dot{e}_2 &= e_3 \\ \dot{e}_3 &= \Psi_L(x_L) - \Psi_C(x_C) - \bar{u}(x_C, y_C) \end{aligned} \tag{2.24}$$

2.4.1 Stability Analysis

If the control input $\bar{u}$ is defined as

$$\bar{u}(x_C, y_L) = \dot{u}_3 = \Psi_L(x_L) - \Psi_C(x_C) + Ke_z$$

with gain vector $\mathrm{K} = [k_1, k_2, k_3]$, then the closed-loop dynamics of the synchronization error in the coordinate transformation is given by $\dot{e}_z = Ae_z$ where $A \in \mathbb{R}^{3\times3}$:

$$A = \begin{bmatrix} 0 & 1 & 0 \\ 0 & 0 & 1 \\ -k_1 & -k_2 & -k_3 \end{bmatrix}$$

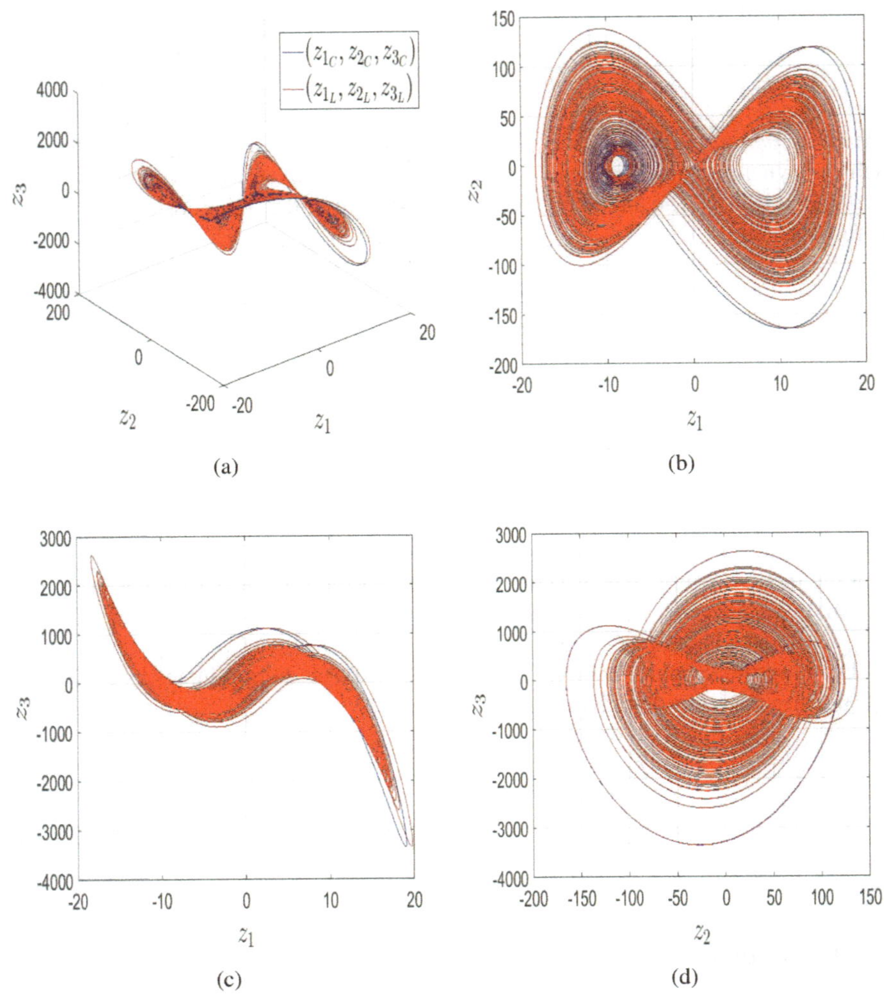

Fig. 2.1 Synchronization of master and slave trajectories in the coordinate transformation (**a**), i.e., $z_L = \Phi_L(x_L)$ and $z_C = \Phi_C(x_C)$. These trajectories approximate each other as $t \to \infty$ by the action of the designed controller. This is clear on the different planes of the phase space (**b**), (**c**), (**d**)

Thus, by means of the Routh–Hurwitz criterion, we conclude that $||e_z|| \to 0$ as $t \to \infty$ if $k_1 > 0$, $k_3 > 0$ and $k_2 > \frac{k_1}{k_3}$.

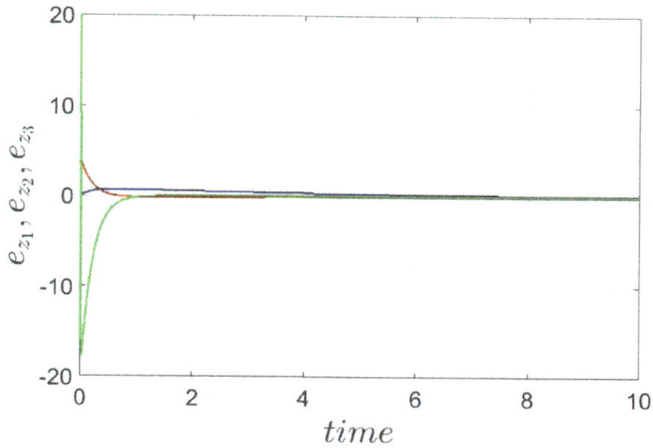

Fig. 2.2 Synchronization errors in the transformed coordinates. It is clear that $||e_z|| \to 0$ as $t \to \infty$. Hence, the proposed approach is effective in order to reach a synchronous state between master and slave systems

2.4.2 Simulation Results

Figure 2.1 shows the synchronization of master and slave systems in the coordinate transformation, with $a_L = 10$, $b_L = 28$, $c_L = 8/3$, $a_C = 15$, $b_C = 25.58$, $m_0 = -5/7$, $m_1 = -8/7$, $k_1 = 130$, $k_2 = 790$ and $k_3 = 130$.

The inverse transformations lead us to

$$\begin{bmatrix} x_{1_L} \\ x_{2_L} \\ x_{3_L} \end{bmatrix} = \begin{bmatrix} z_{1_L} \\ z_{1_L} + \frac{z_{2_L}}{a_L} \\ -\frac{z_{3_L}}{a_L z_{1_L}} - \left(1 + \frac{1}{a_L}\right)\frac{z_{2_L}}{z_{1_L}} + b_L - 1 \end{bmatrix} = \phi_L^{-1}(z_L) \tag{2.25}$$

and

$$\begin{bmatrix} x_{1_C} \\ x_{2_C} \\ x_{3_C} \end{bmatrix} = \begin{bmatrix} -\frac{1}{b_C}\left(z_{3_C} - u_3\right) - \frac{1}{b_C}\left(z_{2_C} - u_2\right) - z_{1_C} + u_1 \\ -\frac{1}{b_C}\left(z_{2_C} - u_2\right) \\ z_{1_C} - u_1 \end{bmatrix} = \phi_C^{-1}(z_C) \tag{2.26}$$

In Fig. 2.2 are given the corresponding synchronization errors in the transformed coordinates, which illustrates the performance of the proposed approach.

Finally, the transformation that satisfies Definition 2.6 is given by,

$$H_{LC} = \phi_L^{-1} \circ \phi_C = \begin{bmatrix} -\frac{1}{b_C}\left(a_L \dot{x}_{2_L} - (a_L - 1)\dot{x}_{1_L}\right) - x_{1_L} + \frac{1}{b_C}(u_2 + u_3) + u_1 \\ -\frac{1}{b_C}\left(\dot{x}_{1_L} - u_2\right) \\ x_{1_L} - u_1 \end{bmatrix}$$

Therefore, the GS state is achieved. Moreover, Fig. 2.3 shows that $||H_{LC}(x_C) - x_L|| \to 0$ as $t \to \infty$. With this example, it is shown that the differential-algebraic approach can synchronize both systems. Note that as an advantage of this approach, the mapping that relates both system trajectories is explicitly found.

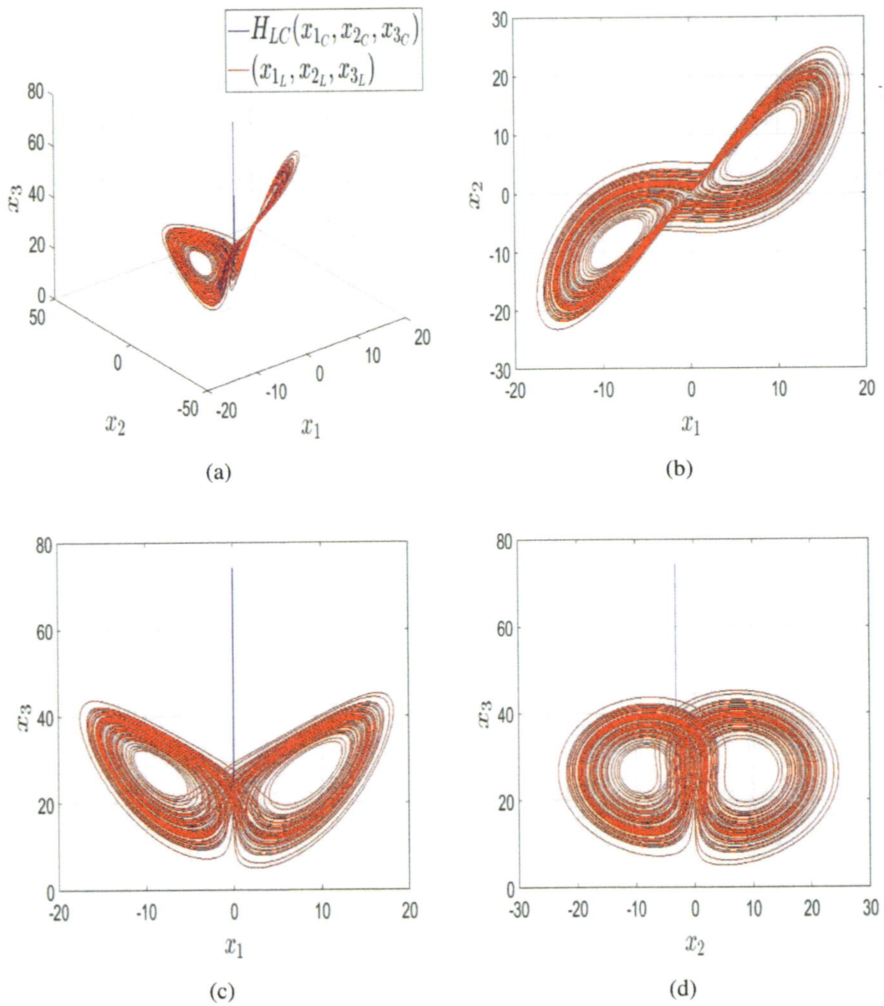

Fig. 2.3 Generalized synchronization of master and slave systems. Through the transformation H_{LC} slave system trajectory and master system trajectory evolve in the same chaotic attractor, in this case, the Lorenz attractor (**a**), i.e., $H_{LC}(x_C) = x_L$. This is clear on the different planes of the phase space (**b**), (**c**), (**d**)

References

1. Abarbanel, H.D., Rulkov, N.F., Sushchik, M.M.: Generalized synchronization of chaos: the auxiliary system approach. Phys. Rev. E **53**(5), 4528 (1996)
2. Dmitriev, B.S., Hramov, A.E., Koronovskii, A.A., Starodubov, A.V., Trubetskov, D.I., Zharkov, Y.D.: First experimental observation of generalized synchronization phenomena in microwave oscillators. Phys. Rev. Lett. **102**(7), 074101 (2009)
3. He, W., Cao, J.: Generalized synchronization of chaotic systems: an auxiliary system approach via matrix measure. Chaos Interdiscip. J. Nonlinear Sci. **19**(1), 013118 (2009)
4. Hramov, A.E., Koronovskii, A.A.: Generalized synchronization: a modified system approach. Phys. Rev. E **71**(6), 067201 (2005)
5. Huang, Y., Wang, Y.-W., Xiao, J.-W.: Generalized lag-synchronization of continuous chaotic system. Chaos, Solitons Fractals **40**(2), 766–770 (2009)
6. Juan, M., Xingyuan, W.: Generalized synchronization via nonlinear control. Chaos Interdiscip. J. Nonlinear Sci. 18(2), 023108 (2008)
7. Kennel, M.B., Brown, R., Abarbanel, H.D.: Determining embedding dimension for phase-space reconstruction using a geometrical construction. Phys. Rev. A **45**(6), 3403 (1992)
8. Kittel, A., Parisi, J., Pyragas, K.: Generalized synchronization of chaos in electronic circuit experiments. Phys. D Nonlinear Phenom. **112**(3–4), 459–471 (1998)
9. Kocarev, L., Parlitz, U.: Generalized synchronization, predictability, and equivalence of unidirectionally coupled dynamical systems. Phys. Rev. Lett. **76**(11), 1816 (1996)
10. Kolchin, E.R.: Differential Algebra & Algebraic Groups, vol. 54. Academic Press (1973)
11. Liu, H., Chen, J., Lu, J.-A., Cao, M.: Generalized synchronization in complex dynamical networks via adaptive couplings. Phys. A Stat. Mech. Its Appl. **389**(8), 1759–1770 (2010)
12. Liu, W., Qian, X., Yang, J., Xiao, J.: Antisynchronization in coupled chaotic oscillators. Phys. Lett. A **354**(1–2), 119–125 (2006)
13. Martínez-Guerra, R., Gómez-Cortés, G., Pérez-Pinacho, C.: Synchronization of integral and fractional order chaotic systems. In: A Differential Algebraic and Differential Geometric Approach. Springer (2015)
14. Martínez-Guerra, R., Pérez-Pinacho, C.A., Gómez-Cortés, G.C.: Generalized synchronization via the differential primitive element. In: Synchronization of Integral and Fractional Order Chaotic Systems, pp. 163–174. Springer (2015)
15. Moskalenko, O.I., Koronovskii, A.A., Hramov, A.E.: Generalized synchronization of chaos for secure communication: remarkable stability to noise. Phys. Lett. A **374**(29), 2925–2931 (2010)
16. Rul'Kov, N., Volkoskii, A., Rodriguez-Lozano, A., Del Rio, E., Velarde, M.: Mutual synchronization of chaotic self-oscillators with dissipative coupling. Int. J. Bifurc. Chaos **2**(03), 669–676 (1992)
17. Rulkov, N.F., Sushchik, M.M., Tsimring, L.S., Abarbanel, H.D.: Generalized synchronization of chaos in directionally coupled chaotic systems. Phys. Rev. E **51**(2), 980 (1995)
18. Sun, M., Zeng, C., Tian, L.: Linear generalized synchronization between two complex networks. Commun. Nonlinear Sci. Numer. Simul. **15**(8), 2162–2167 (2010)
19. Wang, Y.-W., Guan, Z.-H.: Generalized synchronization of continuous chaotic system. Chaos, Solitons Fractals **27**(1), 97–101 (2006)
20. Wang, Y.-W., Wen, C., Yang, M., Xiao, J.-W.: Adaptive control and synchronization for chaotic systems with parametric uncertainties. Phys. Lett. A **372**(14), 2409–2414 (2008)
21. Zhou, J., Chen, J., Lu, J., Lü, J.: On applicability of auxiliary system approach to detect generalized synchronization in complex network. IEEE Trans. Autom. Control **62**(7), 3468–3473 (2016)

Chapter 3
State Estimation and Synchronization

In previous chapters, the synchronization and GS of chaotic systems were described. Nonetheless, these phenomena are not restricted to this particular class of systems. In fact, the unidirectionally coupled dynamic systems

$$\Sigma_m := \begin{cases} \dot{x}_m = F_m(x_m, u_m) \\ y_m = h_m(x_m) \end{cases}$$

and

$$\Sigma_s := \begin{cases} \dot{x}_s = F_s(x_s, u_s(x_s, y_m)) \\ y_s = h_s(x_s) \end{cases}$$

are said to be synchronized if $x_s \to x_m$ as $t \to \infty$. Or equivalently, if

$$\lim_{t\to\infty} ||e(t)|| = 0$$

where $e(t) = x_m - x_s$ is called the synchronization error.

Notice that this definition is ambiguous concerning the class of the dynamic systems Σ_m and Σ_s. Therefore, suppose the particular case where the follower/slave system is an exact or partial copy of the driving/master system, in such a case, that the problem described is a state estimation problem, which is another fundamental topic in control theory.[1]

This chapter demonstrates that synchronization and state estimation can occur simultaneously. Moreover, it is particularly relevant to the usage of the algebraic

[1] This idea was first suggested by Nijmeijer and Mareels [29], who claimed that the similarity between synchronization and state estimation also lies in the fact that in both problems is necessary to design a mechanism such that through a transmitted signal, both systems behave identically.

R. Martínez-Guerra and J. P. Flores-Flores, *An Approach to Multi-agent Systems as a Generalized Multi-synchronization Problem*, Understanding Complex Systems,
https://doi.org/10.1007/978-3-031-22669-4_3

observability condition to design a model-free-based Proportional-Integral reduced-order observer. In addition, the proposed observer is applied to a current real-world problem, the SARS-CoV-2 pandemic.

3.1 Differential Algebra and State Estimation

In control theory, a fundamental notion is the observability property. Most state observers or estimators are designed based on this property. In the context of differential algebra, the equivalent concept is the so-called *algebraic observability condition* (AOC).

First of all, let us recall Definition 2.1 and in what follows, consider the field K as the field of rational numbers, i.e., K $= \mathbb{R}$.

Example 3.1 Consider the following linear system

$$\begin{aligned} \dot{x}_1 &= x_2 + u \\ \dot{x}_2 &= x_1 \\ y &= x_1 \end{aligned}$$

where $x = (x_1, x_2)^T$ is a state vector and y is the output of the system. Then, it is possible to express

$$\begin{aligned} P(x_1) &= y - x_1 = 0 \\ P(x_2) &= \dot{y} - x_2 - u = 0 \end{aligned}$$

with coefficients $u, y, \dot{y}$ in the differential field $\mathbb{R} < u, y >$.

Example 3.2 Now, consider the following nonlinear system

$$\begin{aligned} \dot{x}_1 &= u x_2 x_1 \\ \dot{x}_2 &= u x_1 \\ y &= x_1 \end{aligned}$$

such that

$$\begin{aligned} P(x_1) &= y - x_1 = 0 \\ P(x_2) &= \dot{y} - u x_2 y = 0 \end{aligned}$$

whose coefficients $u, y, \dot{y} \in \mathbb{R} < u, y >$.

Notice from the examples above that $x \in \mathbb{R}^2$ satisfies a differential polynomial. Therefore, the state x is said to be differentially algebraic.

Definition 3.1 Let $x = (x_1, x_2, \ldots, x_n)^T$ be a state vector of a dynamic system. A state variable x_i, $i = 1, 2, \ldots, n$ is said to be *algebraically observable* if satisfies a differential polynomial of the known variables, i.e., $P(x_i) = 0$ with coefficients in $\mathbb{R} < u, y >$. If all state variables of the dynamic system are algebraically observable, it is said that the system satisfies an algebraic observability condition (AOC).

Definition 3.2 A dynamic system is said to be *nondifferentially flat* if any of its variables is not differentially algebraic. That is, the polynomial differential equation $P(x_i, u_i) = 0$ with coefficients in $\mathbb{R} < y >$ is not satisfied for some x_i, $i = 1, 2, \ldots, n$. Otherwise it is said to be *differentially flat*.

Example 3.3 Let us consider the following Lorenz chaotic system

$$\begin{aligned}
\dot{x}_1 &= \sigma(x_2 - x_1) \\
\dot{x}_2 &= \rho x_1 - x_2 - x_1 x_3 \\
\dot{x}_3 &= x_1 x_2 - \beta x_3 \\
y &= x_1
\end{aligned}$$

and after some algebraic manipulation

$$\begin{aligned}
P(x_1) &= y - x_1 = 0 \\
P(x_2) &= \sigma(x_2 - y) - \dot{y} = 0 \\
P(x_3) &= y(x_3 - \rho + 1) + \dot{y}\left(1 + \frac{1}{\sigma}\right) + \frac{\ddot{y}}{\sigma} = 0
\end{aligned}$$

It is possible to note that $P(x_3)$ is not rational since $y \neq 0$, and its derivatives are coefficients in $\mathbb{R} < y >$. Therefore, the system is said to be differentially flat.

Example 3.4 Consider once more the Lorenz chaotic system, now with the following output signal $y = x_2$

$$\begin{aligned}
P(x_1, \dot{x}_1) &= \dot{x}_1 - \sigma(y - x_1) = 0 \\
P(x_2) &= y - x_2 = 0 \\
P(x_1, x_3, \dot{x}_3) &= \dot{x}_3 + \beta x_3 - y x_1 = 0
\end{aligned}$$

In this case, the system is nondifferentially flat, since Definition 3.2 is not satisfied.

Remark 3.1 A nondifferentially flat system does not fulfill the AOC. Nonetheless, under some considerations, it is still possible to design an observer for this class of systems. The result is known as a reduced-order state observer.

3.2 Reduced-Order PI Observer

Let us consider the following dynamical nonlinear system

$$\begin{aligned} \dot{x} &= F(x, u) \\ y &= h(x) \end{aligned} \tag{3.1}$$

where $x = (x_1, x_2, ..., x_n)^T$ is the state vector of the system, which is not fully available, F is an analytic nonlinear function, and h is a polynomial of its arguments. Variables $u = (u_1, u_2, ..., u_m)^T$ and $y \in \mathbb{R}$ are the input and output signals, respectively. The aim is to estimate the unknown variables from the available measurements, such that in the following is presented a state observer or estimator.[2]

Consider the auxiliary variable η, defined as a function of the unknown variables,[3] such that, system (3.1) can be expressed as the following immersion system (see [8])

$$\begin{aligned} \dot{x}(t) &= f(x, \eta, u) \\ \dot{\eta}(x) &= \Omega(x) \\ y(t) &= h(x) \end{aligned} \tag{3.2}$$

We can note that in the new system's representation the unknown variable is η, such that the problem now is to estimate the value of η. That is, a state observer has to be designed for this auxiliary variable. Thus, consider the following conditions:

H1 The auxiliary variable η is algebraically observable.
H2 The unknown dynamics $\Omega(x)$ is bounded, that is, $||\Omega(x)|| \leq N$, where $N > 0$.
H3 There exist $\gamma_1, \gamma_2 \in C^1$ real-valued functions.

Theorem 3.1 *If H1–H3 are fulfilled, then*

$$\begin{aligned} \dot{\hat{\eta}} &= k_p(\eta - \hat{\eta}) + \hat{\eta}_1 \\ \dot{\hat{\eta}}_1 &= k_i(\eta - \hat{\eta}) \end{aligned} \tag{3.3}$$

is a model-free-based[4] *Proportional-Integral reduced-order observer for system (3.2), where $\hat{\eta}$ is the estimation of η and $\hat{\eta}_1$ is the integral part of the observer, whose convergence error defined as*

---

[2] The vast majority of control algorithms require knowing all or at least some state variables. In plenty of cases, it is possible to implement sensors. However, there are exceptions due to economic or technological restrictions. Therefore, the idea of designing a mathematical algorithm for obtaining such unknown variables stands out.

[3] The selection of η is not unique and depending on how is chosen, the variables of interest are determined directly or indirectly. On the other hand, it could be the case that η is defined as $\eta = [\eta_1, ..., \eta_\mu]$, such that, a bank of observers has to be designed, that is, a state observer for each η_i.

[4] This observer is called model-free since an exact copy of the dynamical system is not required. This is a relevant property of this methodology since represents a more natural methodology.

$$e = \eta - \hat{\eta} \tag{3.4}$$

is uniformly ultimately bounded.

Proof System (3.3) can be expressed in the following way

$$\dot{\hat{\eta}} = k_p e_1 + k_i e_2 \tag{3.5}$$

where $e_1 = \eta - \hat{\eta}$ and $\dot{e}_2 = e_1$. On the other hand, let E be

$$E = \begin{pmatrix} e_1 \\ e_2 \end{pmatrix} \tag{3.6}$$

whose derivative is

$$\begin{aligned}\dot{E} &= \begin{pmatrix} \dot{\eta} - \dot{\hat{\eta}} \\ e_1 \end{pmatrix} \\ &= \begin{pmatrix} \Omega - k_p e_1 - k_i e_2 \\ e_1 \end{pmatrix} \\ &= -\begin{pmatrix} k_p & k_i \\ -1 & 0 \end{pmatrix}\begin{pmatrix} e_1 \\ e_2 \end{pmatrix} + \begin{pmatrix} \Omega \\ 0 \end{pmatrix}\end{aligned} \tag{3.7}$$

or in matrix form

$$\dot{E} = -\mathrm{K}E + \bar{\Omega} \tag{3.8}$$

Now, let us consider the following Lyapunov candidate function

$$V(E) = E^T P E \tag{3.9}$$

with $P = I$. Then, we have that the derivative of $V(E)$ along the trajectories of (3.8) is

$$\begin{aligned}\dot{V}(E) &= \dot{E}^T E + E^T \dot{E} \\ &= 2E^T \dot{E} \\ &= 2E^T \left(-\mathrm{K}E + \bar{\Omega}\right) \\ &= -2E^T \mathrm{K}E + 2E^T \bar{\Omega}\end{aligned} \tag{3.10}$$

Moreover, from Rayleigh inequality, we have that the first term of $\dot{V}(E)$ is bounded, i.e.,

$$-2E^T \mathrm{K}E \le -2\lambda_{\min}(\mathrm{K})||E||^2 \tag{3.11}$$

and, from H2, $\bar{\Omega}$ is such that

$$||2E^T\bar{\Omega}|| \leq 2N||E||_{\mathrm{K}} = 2N\sqrt{\lambda_{\max}(\mathrm{K})}||E|| \tag{3.12}$$

Therefore, we conclude that

$$\dot{V}(E) \leq -2\lambda_{\min}(\mathrm{K})||E||^2 + 2N\sqrt{\lambda_{\max}(\mathrm{K})}||E|| \tag{3.13}$$

Finally, by applying the uniformly ultimately boundedness theorem [10], it directly follows that $E(t)$ is bounded uniformly for any initial condition $E(0)$, and $E(t)$ remains in the compact set $\mathscr{B}_\delta = \{E : ||E|| \leq \delta, \delta > 0\}$. Where

$$\delta = \sqrt{\frac{\lambda_{\max}(\mathrm{K})}{\lambda_{\min}(\mathrm{K})}\left(\frac{2N\sqrt{\lambda_{\max}(\mathrm{K})}}{\lambda_{\min}(\mathrm{K})}\right)} > 0 \tag{3.14}$$

Finally, the selection of gains k_p and k_i can be derived from expression (3.14). That is, consider

$$\frac{\lambda_{\max}(\mathrm{K})}{\lambda_{\min}(\mathrm{K})}\frac{4N^2\lambda_{\max}(\mathrm{K})}{\lambda_{\min}^2(\mathrm{K})} = \frac{4N^2\lambda_{\max}^2(\mathrm{K})}{\lambda_{\min}^3(\mathrm{K})} > 0 \tag{3.15}$$

Then, to satisfy the last inequality, $\lambda_{\min}(\mathrm{K}) > 0$ and, therefore, $\lambda_{\max}(\mathrm{K}) > 0$. □

3.2.1 Removing Dependence of Output's Derivatives

Sometimes, depending on the choice of the auxiliary variable η, the derivatives of the output signal might be required. These derivatives are usually unavailable. Therefore, consider the methodology to avoid them.

Lemma 3.1 *If the auxiliary variable η_i, $i \in \{1, ..., \mu\}$, satisfies H1 and can be written as*

$$\eta_i = a_i\dot{y} + b_i(u, y) \tag{3.16}$$

where a_i is constant and $b_i(u, y)$ is a bounded function, then there exist functions γ_{1_i} and $\gamma_{2_i} \in C^1$ such that the model-free-based PI reduced-order observer (3.3) *can be expressed as*

$$\begin{aligned}
\dot{\gamma}_{1_i} &= -k_{p_i}\gamma_{1_i} + \gamma_{2_i} + k_{p_i}\left[b_i(u, y) - k_{p_i}a_i y\right] + k_{i_i}a_i y,\\
\gamma_{1_i}(0) &= \gamma_{1i_0}\\
\dot{\gamma}_{2_i} &= -k_{i_i}\gamma_{1_i} + k_{i_i}\left[b_i(u, y) - k_{p_i}a_i y\right], \quad \gamma_{2_i}(0) = \gamma_{2i_0}\\
\hat{\eta}_i &= \gamma_{1_i} + k_{p_i}a_i y\\
\hat{\eta}_{1_i} &= \gamma_{2_i} + k_{i_i}a_i y
\end{aligned} \tag{3.17}$$

Proof Let us consider the model-free-based PI reduced-order observer

$$\begin{aligned}
\dot{\hat{\eta}}_i &= k_{p_i}(\eta_i - \hat{\eta}_i) + \hat{\eta}_{1_i}\\
\dot{\hat{\eta}}_{1_i} &= k_{i_i}(\eta_i - \hat{\eta}_i),
\end{aligned} \tag{3.18}$$

$i = 1, 2, \ldots, \mu$. Substituting (3.16) in (3.18),

$$\begin{aligned}
\dot{\hat{\eta}}_i &= k_{p_i}(a_i\dot{y} + b_i(u, y) - \hat{\eta}_i) + \hat{\eta}_{1_i}\\
\dot{\hat{\eta}}_{1_i} &= k_{i_i}(a_i\dot{y} + b_i(u, y) - \hat{\eta}_i)
\end{aligned} \tag{3.19}$$

Then, let the artificial variables γ_{1_i} and $\gamma_{2_i} \in C^1$ be

$$\begin{aligned}
\gamma_{1_i} &= \hat{\eta}_i - k_{p_i}a_i y\\
\gamma_{2_i} &= \hat{\eta}_{1_i} - k_{i_i}a_i y
\end{aligned} \tag{3.20}$$

whose derivatives are

$$\begin{aligned}
\dot{\gamma}_{1_i} &= \dot{\hat{\eta}}_i - k_{p_i}a_i\dot{y}\\
\dot{\gamma}_{2_i} &= \dot{\hat{\eta}}_{1_i} - k_{i_i}a_i\dot{y}
\end{aligned} \tag{3.21}$$

Finally by considering (3.19)–(3.21),

$$\begin{aligned}
\dot{\gamma}_{1_i} &= -k_{p_i}\gamma_{1_i} + \gamma_{2_i} + k_{p_i}\left[b_i(u, y) - k_{p_i}a_i y\right] + k_{i_i}a_i y,\\
\gamma_{1_i}(0) &= \gamma_{1i_0}\\
\dot{\gamma}_{2_i} &= -k_{i_i}\gamma_{1_i} + k_{i_i}\left[b_i(u, y) - k_{p_i}a_i y\right],\\
\gamma_{2_i}(0) &= \gamma_{2i_0}
\end{aligned} \tag{3.22}$$

and the proof is completed. □

3.2.2 *PI Observer: Numerical Examples*

In the following, we present some implementation examples of the model-free reduced-order PI observer.

Example 3.5 Consider the following linear system

$$\begin{aligned}\dot{x}_1 &= u \\ \dot{x}_2 &= u + f \\ y &= x_2\end{aligned} \tag{3.23}$$

where u is a known input signal and f is and unknown variable. We can verify that system (3.23) is nondifferentially flat, since

$$\begin{aligned}P(\dot{x}_1) &= \dot{x}_1 - u = 0 \\ P(x_2) &= x_2 - y = 0 \\ P(f) &= u + f - \dot{y} = 0\end{aligned} \tag{3.24}$$

Moreover, the linear system can be expressed as

$$\begin{aligned}\dot{x}_1 &= u \\ \dot{x}_2 &= u + f \\ \dot{f} &= \Omega(x_1, x_2) \\ y &= x_2\end{aligned} \tag{3.25}$$

Thus, the PI observer is expressed as the following set of differential equations

$$\begin{aligned}\dot{\hat{f}} &= k_p(f - \hat{f}) + \hat{f}_1 \\ \dot{\hat{f}}_1 &= k_i(f - \hat{f})\end{aligned} \tag{3.26}$$

where $\hat{f}$ is the estimate of f and $\hat{f}_1$ represents the integral part of the PI observer. It is clear that f is algebraically observable, since $f = \dot{y} - u$. On the other hand, let us suppose that $\dot{f}$ is bounded. Then, we have

$$\begin{aligned}\dot{\hat{f}} &= k_p(\dot{y} - u - \hat{f}) + \hat{f}_1 \\ \dot{\hat{f}}_1 &= k_i(\dot{y} - u - \hat{f})\end{aligned} \tag{3.27}$$

Notice that $\dot{y}$ is involved in the last set of equations. Since $\dot{y}$ is not available, we introduce the following artificial variables

$$\begin{aligned}\gamma_1 &= \hat{f} - k_p y \\ \gamma_2 &= \hat{f}_1 - k_i y\end{aligned} \tag{3.28}$$

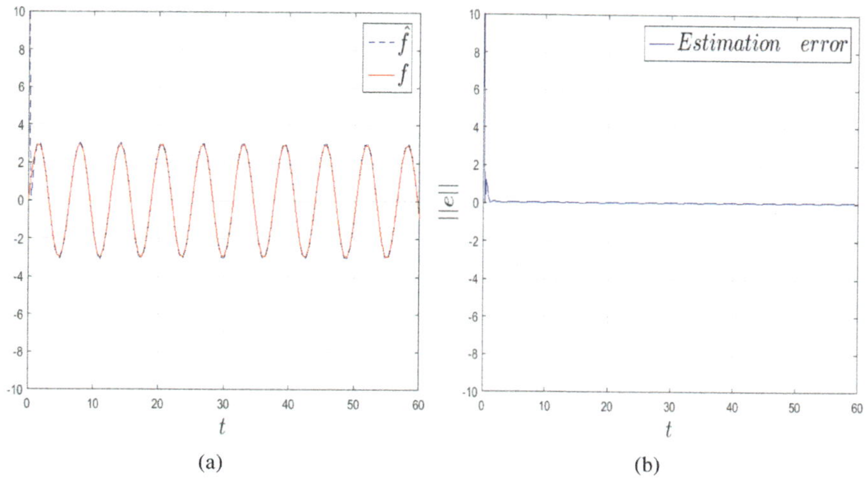

Fig. 3.1 **a** Estimation of f and **b** norm of the corresponding estimation error obtained by the model-free PI reduced-order observer

By deriving the artificial variables and replacing $\hat{f}$ and $\hat{f}_1$, we have

$$\begin{aligned}\dot{\gamma}_1 &= -k_p(u + \gamma_1 + k_p y) + \gamma_2 + k_i y \\ \dot{\gamma}_2 &= -k_i u - k_i(\gamma_1 + k_p y)\end{aligned} \tag{3.29}$$

Such that the model-free-based PI observer is given by

$$\begin{aligned}\dot{\gamma}_1 &= -k_p\gamma_1 + \gamma_2 - k_p(u + k_p y) + k_i y, \qquad &\gamma_1(0) = \gamma_{1_0} \\ \dot{\gamma}_2 &= -k_i\gamma_1 - k_i(u + k_p y), \qquad &\gamma_2(0) = \gamma_{2_0} \\ \hat{f} &= \gamma_1 + k_p y \\ \hat{f}_1 &= \gamma_2 + k_i y\end{aligned} \tag{3.30}$$

Consider the following simulation parameters: $\gamma_1(0) = 0$, $\gamma_2(0) = 0$, $u = 2\sin(\pi t)$, $f = 3\sin(\pi t)$ and the following gains: $k_i = 15$. $k_p = 8$. The numerical results are shown in Fig. 3.1. Notice how the estimation error tends to zero as $t \to \infty$, such that the variable of interest f is correctly estimated.

Example 3.6 Let us consider the Chua chaotic system attractor, described by the next set of nonlinear differential equations[5]

[5] This system represents a simple electronic circuit designed by Leon O. Chua. Despite its simplicity, this circuit exhibits chaotic behavior and was the first system of this kind to be proven by analytical and experimental means. It is constituted of two capacitors and one inductor [5].

$$\begin{aligned}
\dot{x}_1 &= a(x_2 - x_1 - m_0 x_1 - x_1^3 m_1) \\
\dot{x}_2 &= x_1 - x_2 + x_3 \\
\dot{x}_3 &= -b x_2 \\
y &= x_2
\end{aligned} \tag{3.31}$$

where a, b, m_0 and m_1 are constant parameters. The variables x_1 and x_2 are the voltage across each capacitor; meanwhile, x_3 is the electric current in the inductor.

System (3.31) is nondifferentially flat, since

$$\begin{aligned}
P(x_1, x_3) &= x_1 - \dot{y} - y + x_3 = 0 \\
P(x_2) &= x_2 - y = 0 \\
P(\dot{x}_3) &= \dot{x}_3 + by = 0
\end{aligned} \tag{3.32}$$

Therefore, we define the variable

$$\eta = x_1 + x_3 = \dot{y} + y \tag{3.33}$$

which satisfies H1. Thus, system (3.31) is expressed as

$$\begin{aligned}
\dot{x}_1 &= a(x_2 - x_1 - m_0 x_1 - x_1^3 m_1) \\
\dot{x}_2 &= -x_2 + \eta \\
\dot{x}_3 &= -b x_2 \\
\dot{\eta} &= \Omega(x_1, x_3) \\
y &= x_2
\end{aligned} \tag{3.34}$$

The Chua system is chaotic; therefore, the dynamics $\Omega(x_1, x_3)$ satisfies H2. The PI observer for this case is given by

$$\begin{aligned}
\dot{\hat{\eta}} &= k_p(\dot{y} + y - \hat{\eta}) + \hat{\eta}_1 \\
\dot{\hat{\eta}}_1 &= k_i(\dot{y} + y - \hat{\eta})
\end{aligned} \tag{3.35}$$

Then, to remove the dependency on $\dot{y}$, let

$$\begin{aligned}
\gamma_1 &= \hat{\eta} - k_p y \\
\gamma_2 &= \hat{\eta}_1 - k_i y
\end{aligned} \tag{3.36}$$

such that

$$\begin{aligned}
\dot{\gamma}_1 &= k_p(y - \gamma_1 - k_p y) + \gamma_2 + k_i y \\
\dot{\gamma}_2 &= k_i(y - \gamma_1 - k_p y)
\end{aligned} \tag{3.37}$$

And therefore, the model-free-based PI observer is given by

$$\begin{aligned}
\dot{\gamma}_1 &= -k_p\gamma_1 + \gamma_2 + k_p(y - k_p y) + k_i y, & \gamma_1(0) &= \gamma_{1_0} \\
\dot{\gamma}_2 &= -k_i\gamma_1 + k_i(y - k_p y), & \gamma_2(0) &= \gamma_{2_0} \\
\hat{\eta} &= \gamma_1 + k_p y \\
\hat{\eta}_1 &= \gamma_2 + k_i y
\end{aligned} \tag{3.38}$$

The estimated value $\hat{\eta}$ can be used to determine x_1 and x_3 as follows: From expression (3.33), consider the following relationships

$$\hat{\eta} = \hat{x}_1 + x_3 \tag{3.39}$$

and

$$\dot{y} = \hat{\eta} - y \tag{3.40}$$

Deriving (3.39), it is obtained

$$\dot{\hat{\eta}} = \dot{\hat{x}}_1 + \dot{x}_3 = \dot{\hat{x}}_1 - by \tag{3.41}$$

On the other hand

$$\hat{\eta} = \gamma_1 + k_p y \tag{3.42}$$

such that

$$\dot{\hat{\eta}} = \dot{\gamma}_1 + k_p\dot{y} \tag{3.43}$$

Then, from (3.41) and (3.43)

$$\dot{\hat{x}}_1 = \dot{\gamma}_1 + k_p\dot{y} + by \tag{3.44}$$

and substituting $\dot{\gamma}_1$ from (3.38) and (3.40) in (3.44),

$$\begin{aligned}
\dot{\hat{x}}_1 = &- k_p\gamma_1 + \gamma_2 + k_p(y - k_p y) \\
&+ k_i y + k_p(\hat{\eta} - y) + by
\end{aligned} \tag{3.45}$$

Thus, by substituting (3.42) in (3.45) and simplifying

$$\begin{aligned}
\dot{\hat{x}}_1 &= \gamma_2 + k_i y + by \\
&= \hat{\eta}_1 + by
\end{aligned} \tag{3.46}$$

The last equation (3.46) can be solved with the method of Euler, i.e.,

$$\begin{aligned}\hat{x}_1(t+1) &= \hat{x}_1(t) + h[\hat{\eta}_1(t) + by(t)] \\ \hat{x}_1(0) &= \hat{x}_{1_0}\end{aligned} \tag{3.47}$$

and finally, from (3.39)

$$\hat{x}_3 = \hat{\eta} - \hat{x}_1 \tag{3.48}$$

such that all variables of interest can be estimated. Let the system parameters be $a = 9.5$, $b = 100/7$, $m_0 = -8/7$ and $m_1 = 4/63$, and the initial conditions $x_0 = (0.2, 0.2, 0.2)$, $\gamma_1(0) = 0$, $\gamma_2(0) = 0$ and $\hat{x}_1(0) = 0$. Meanwhile, let the observer gains be $k_i = 150$ and $k_p = 25$. The numerical results can be observed in Fig. 3.2. One can note how the dynamics of the state observer approaches and stays within Chua attractor as $t \to \infty$. In other words, Fig. 3.2 exemplifies better how state estimation and synchronization occurs simultaneously.

Example 3.7 Consider the following nonlinear system

$$\begin{aligned}\dot{x}_1 &= x_2 \\ \dot{x}_2 &= g - \frac{\sigma\alpha u}{x_3} \\ \dot{x}_3 &= -\sigma u - f \\ y &= x_1\end{aligned} \tag{3.49}$$

This system describes the vertical landing of a ship on a planet's surface. This model assumes a constant gravitational acceleration g and no atmospheric resistance. The variables x_1, x_2 and x_3 are the vertical position, descent velocity and mass of the ship, respectively. Meanwhile, u is a control signal, σ is the relative velocity of ejection, and α is a positive constant such that $\sigma\alpha$ is the maximal displacement of the motor breaking. The variable f is an unknown input signal that is considered to be bounded. Notice that this system is nondifferential, since

$$\begin{aligned}P(x_1) &= x_1 - y = 0 \\ P(x_2) &= x_2 - \dot{y} = 0 \\ P(x_3, \dot{x}_3) &= (\ddot{y} - g)x_3 - (\dot{x}_3 + f)\alpha = 0\end{aligned} \tag{3.50}$$

Moreover, it can be expressed as

$$\begin{aligned}\dot{x}_1 &= x_2 \\ \dot{x}_2 &= g - \frac{\sigma\alpha u}{x_3} \\ \dot{x}_3 &= -\sigma u - f \\ \dot{f} &= \Omega(x_1, x_2, x_3) \\ y &= x_1\end{aligned} \tag{3.51}$$

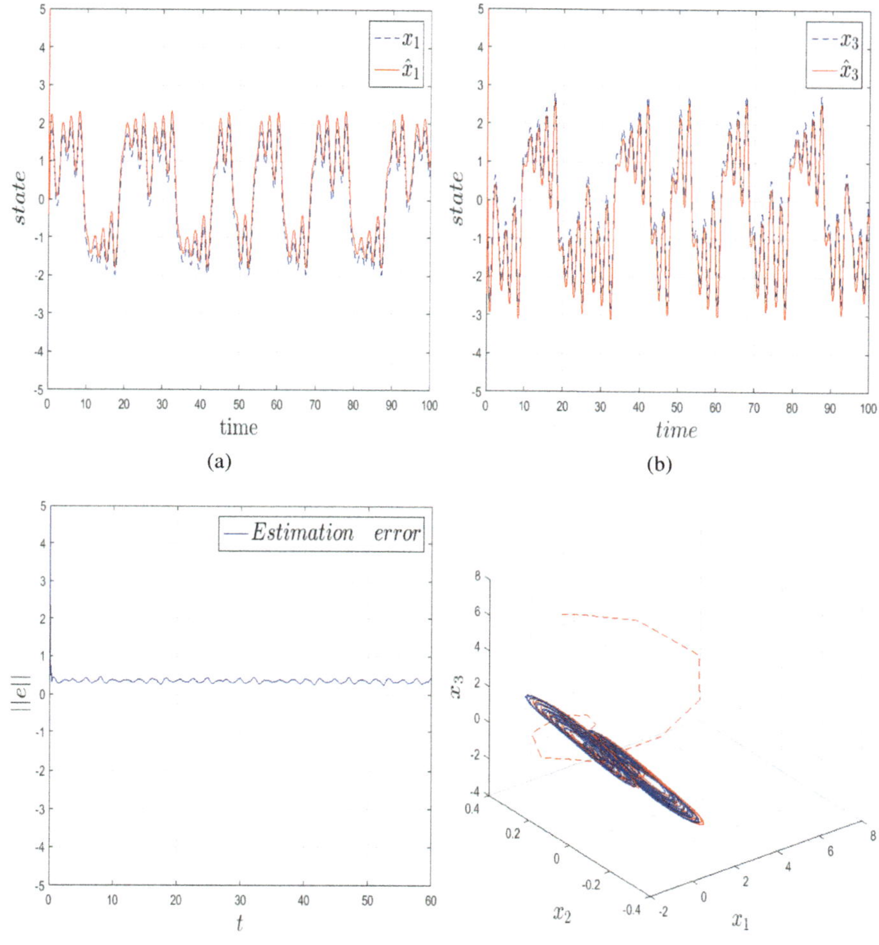

Fig. 3.2 Estimations of **a** x_1 and **b** x_3, **c** resulting estimation error and **d** phase space where the Chua attractor can be observed

One can note that the unknown variable f satisfies H1, since

$$(f + \sigma u)(\ddot{y} - g)^2 + \sigma \alpha u y^{(3)} = 0 \tag{3.52}$$

Then, consider the next PI observer

$$\begin{aligned} \dot{\hat{f}} &= k_{pf}(f - \hat{f}) + \hat{f}_1 \\ \dot{\hat{f}}_1 &= k_{if}(f - \hat{f}) \end{aligned} \tag{3.53}$$

By substituting (3.52),

$$\begin{aligned}\dot{\hat{f}} &= -k_{pf}\left(\sigma u + \frac{\sigma\alpha u y^{(3)}}{(\ddot{y}-g)^2} + \hat{f}\right) + \hat{f}_1 \\ \dot{\hat{f}}_1 &= -k_{if}\left(\sigma u + \frac{\sigma\alpha u y^{(3)}}{(\ddot{y}-g)^2} + \hat{f}\right)\end{aligned} \tag{3.54}$$

Since the variables $y^{(3)}$ and $\ddot{y}$ we approximate them as $y^{(3)} \approx \ddot{\hat{x}}_2$ and $\ddot{y} = \dot{\hat{x}}_2$. Therefore

$$\begin{aligned}\dot{\hat{f}} &= -k_{pf}\left(\sigma u + \frac{\sigma\alpha u \ddot{\hat{x}}_2}{(\dot{\hat{x}}_2-g)^2} + \hat{f}\right) + \hat{f}_1 \\ \dot{\hat{f}}_1 &= -k_{if}\left(\sigma u + \frac{\sigma\alpha u \ddot{\hat{x}}_2}{(\dot{\hat{x}}_2-g)^2} + \hat{f}\right)\end{aligned} \tag{3.55}$$

Now, let γ_{1_f} and γ_{2_f} be artificial variables satisfying H3

$$\begin{aligned}\gamma_{1_f} &= \hat{f} - k_{pf}\frac{\sigma\alpha u}{\dot{\hat{x}}_2 - g} \\ \gamma_{2_f} &= \hat{f}_1 - k_{if}\frac{\sigma\alpha u}{\dot{\hat{x}}_2 - g}\end{aligned} \tag{3.56}$$

Such that their respective derivatives are

$$\begin{aligned}\dot{\gamma}_{1_f} &= \dot{\hat{f}} + k_{pf}\ddot{\hat{x}}_2\frac{\sigma\alpha u}{(\dot{\hat{x}}_2 - g)^2} \\ \dot{\gamma}_{2_f} &= \dot{\hat{f}}_1 + k_{if}\ddot{\hat{x}}_2\frac{\sigma\alpha u}{(\dot{\hat{x}}_2 - g)^2}\end{aligned} \tag{3.57}$$

and by substituting (3.55) and (3.56) in (3.57), the PI observer is given by

$$\begin{aligned}\dot{\gamma}_{1_f} &= -k_{pf}\left(\sigma u + \gamma_{1_f} + k_{pf}\frac{\sigma\alpha u}{\dot{\hat{x}}_2 - g}\right) + \gamma_{2_f} + k_{if}\frac{\sigma\alpha u}{\dot{\hat{x}}_2 - g}, \\ \gamma_{1_f}(0) &= \gamma_{1_f0} \\ \dot{\gamma}_{2_f} &= -k_{if}\left(\sigma u + \gamma_{1_f} + k_{pf}\frac{\sigma\alpha u}{\dot{\hat{x}}_2 - g}\right), \quad \gamma_{2_f}(0) = \gamma_{2_f0} \\ \hat{f} &= \gamma_{1_f} + k_{pf}\frac{\sigma\alpha u}{\dot{\hat{x}}_2 - g} \\ \hat{f}_1 &= \gamma_{2_f} + k_{if}\frac{\sigma\alpha u}{\dot{\hat{x}}_2 - g}\end{aligned} \tag{3.58}$$

Since the variable $\dot{\hat{x}}_2$ is needed, a second PI observer is considered for x_2, i.e.,

$$\begin{aligned}\dot{\hat{x}}_2 &= k_{px_2}(\dot{y} - \hat{x}_2) + \hat{x}_{2_1} \\ \dot{\hat{x}}_{2_1} &= k_{ix_2}(\dot{y} - \hat{x}_2)\end{aligned} \tag{3.59}$$

where $x_2 = \dot{y}$. Thus, let us define the following artificial variables

$$\begin{aligned}\gamma_{1_{x2}} &= \hat{x}_2 - k_{px_2} y \\ \gamma_{2_{x2}} &= \hat{x}_{2_1} - k_{ix_2} y\end{aligned} \tag{3.60}$$

whose derivatives are

$$\begin{aligned}\dot{\gamma}_{1_{x2}} &= \dot{\hat{x}}_2 - k_{px_2} \dot{y} \\ \dot{\gamma}_{2_{x2}} &= \dot{\hat{x}}_{2_1} - k_{ix_2} \dot{y}\end{aligned} \tag{3.61}$$

and by substituting (3.59) and (3.60) in (3.61),

$$\begin{aligned}\dot{\gamma}_{1_{x2}} &= -k_{px_2}\left(\gamma_{1_{x2}} + k_{px_2} y\right) + \gamma_{2_{x2}} + k_{ix_2} y, \\ \gamma_{1_{x2}}(0) &= \gamma_{1_{x2_0}} \\ \dot{\gamma}_{2_{x2}} &= -k_{ix_2}\left(\gamma_{1_{x2}} + k_{px_2} y\right), \quad \gamma_{2_{x2}}(0) = \gamma_{2_{x2_0}} \\ \hat{x}_2 &= \gamma_{1_{x2}} + k_{px_2} y \\ \hat{x}_{2_1} &= \gamma_{2_{x2}} + k_{ix_2} y\end{aligned} \tag{3.62}$$

Finally, since $\dot{y} \approx \hat{x}_2$, from (3.61),

$$\begin{aligned}\dot{\hat{x}}_2 = &- k_{px_2}\left(\gamma_{1_{x2}} + k_{px_2} y\right) \\ &+ \gamma_{2_{x2}} + k_{ix_2} y + k_{px_2} \hat{x}_2\end{aligned} \tag{3.63}$$

such that from the estimation of $\dot{\hat{x}}_2$, the variable of interest f can be obtained.

Notice that for this specific example, a bank of observers was obtained, i.e., a set of PI observers, given by expressions (3.58), (3.62) and (3.63), is used to obtain the variables of interest.

Finally, let $g = 1.63m/s^2, \sigma = 50kg/s, \alpha = 200m/s$ and $x_0 = (-700, 0, 1500)$. On the other hand, $\gamma_{1_{x2}}(0) = 0$, $\gamma_{2_{x2}}(0) = 0$, $\gamma_{1_f}(0)$ and $\gamma_{2_f}(0) = 0$, with $k_{pf} = 49$, $k_{if} = 700$, $k_{px_2} = 6$ and $k_{ix_2} = 25$. Menawhile, the control signal is $u = 1$ and the uncertain variable $f = \exp(6\sin(t))$. The corresponding numerical results can be observed in Fig. 3.3. From this, and the last examples, it is clear that the reduced-order PI observer is capable of correctly estimating the unknown state variables.

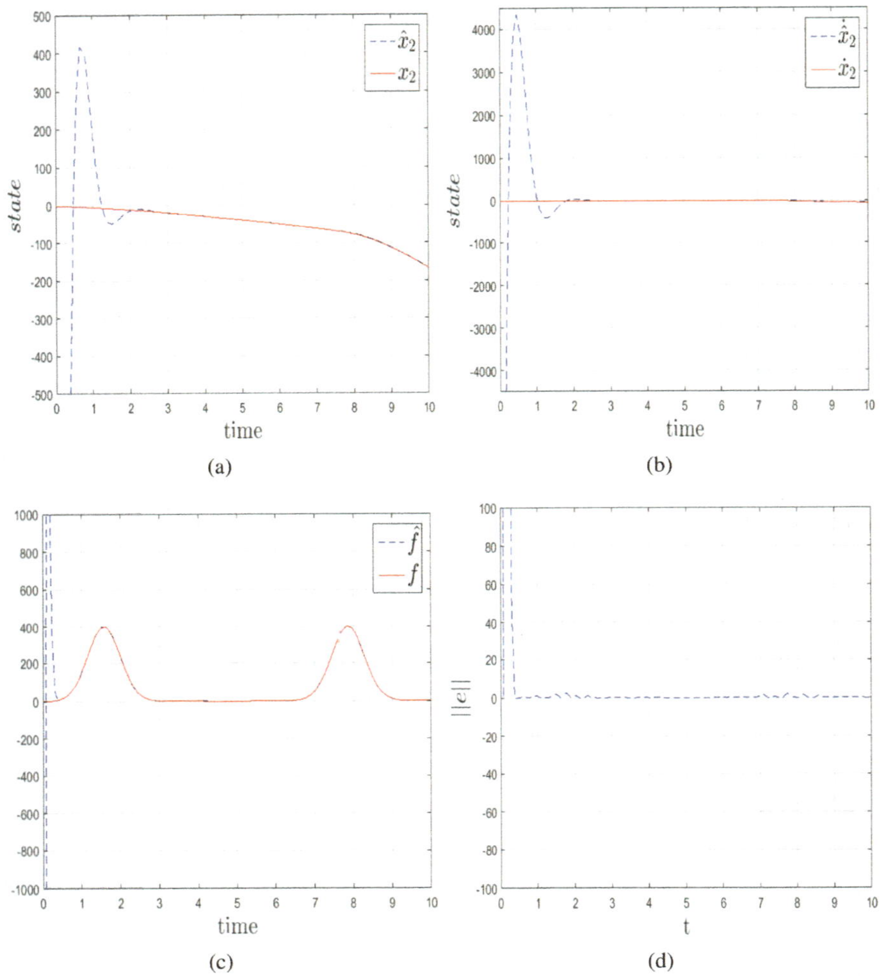

Fig. 3.3 **a** Estimations of the state variable x_2 and **b** its derivative, **c** variable of interest f and **d** corresponding estimation error

3.3 PI Observer: A Pandemic Application

Since 2019, the severe acute respiratory syndrome SARS-CoV-2 virus has overwhelmed every nation due to: (1) its considerable mortality rate, placed between 3–7% during the first months of the pandemic [34, 38] and (2) its basic reproduction number.[6] In addition to these aspects, evidence suggests a considerable presence of

[6] The basic reproduction number is a measure of how contagious a virus is. In other words, it indicates the number of people that an infected individual can transmit the virus [12, 25, 41, 45].

asymptomatic individuals, which are as contagious as the symptomatic carriers [3, 13, 46] and, moreover, are particularly dangerous since they show no reason to take special cautions around them.

To mitigate the pandemic's effects, strict politics, such as massive viral testing campaigns and nonpharmaceutical interventions, was widely promoted by international organizations and several governments around the world. However, for many nations it was impossible to conduct such strategies, due to economic, medical and social conditions [4]. Therefore, as an alternative, experts suggested employing mathematical models for understanding the pandemic's course and, moreover, to design effective mitigation strategies.[7]

The classical mathematical models are the SIR model (Susceptible, Infected and Removed) and the SEIR model (Susceptible, Exposed, Infected and Removed) [9, 16, 37]. However, we can find models much more complex that take into account weather conditions or more specific populations, e.g., weather conditions[8] or hospitalized individuals [2, 21, 28, 33, 47]. Although complex models incorporate a great quantity of biological and epidemiological information, in many cases these involved too many parameters that can be hard to fit. Therefore, here is considered a simple, but still, accurate enough model.

3.3.1 A-SIR Model

A crucial factor of the COVID-19 pandemic is the presence of asymptomatic individuals, which are as contagious as symptomatic ones. The vast majority of asymptomatic individuals remain undetected by most healthcare surveillance systems since they do not require medical attention. Therefore, let us consider the following SIR-type model, known as A-SIR model [14]:

$$
\begin{aligned}
\dot{S}(t) &= -\beta S(t)\left[A(t) + I(t)\right], \\
\dot{I}(t) &= \rho\beta S(t)\left[A(t) + I(t)\right] - \gamma I(t), \\
\dot{A}(t) &= (1-\rho)\,\beta S(t)\left[A(t) + I(t)\right] - \nu A(t), \\
\dot{R}(t) &= \gamma I(t) + \nu A(t), \\
Y(t) &= I(t),
\end{aligned}
\tag{3.64}
$$

Initially, the World Health Organization reported a reproduction number of around 2. However, some other sources quickly reported values of up to 4 and 6 [23, 43].

[7] The idea of employing state observers in the epidemiological context is not new, and in fact, still it is a very active field. These have proved to be particularly useful for designing population restriction policies, finding nonmeasured variables and fitting model parameters [7, 24, 32, 39].

[8] It is well known that any pandemic is highly influenced by external conditions such as air pollution [42], wind velocity, daily sunlight, air pressure, temperature and humidity [17, 18, 36] and even noise pollution [11]. In particular, the COVID-19 pandemic has shown to be sensitive to cold and dry conditions [6].

where $S(t)$ and $R(t)$ represent the populations of susceptible and removed individuals, respectively. The infected population is divided into symptomatic individuals $I(t)$ and asymptomatic individuals $A(t)$.[9] The system's output is $Y(t)$. We suppose that the available data is only the number of symptomatic individuals, i.e., those that have been reported to authorities. The initial conditions of model (3.64) are: $S(0) = S_0$, $I(0) = I_0$, $A(0) = A_0$ and $R(0) = R_0$.

The parameter $\beta > 0$ is known as the disease transmission rate and is given by $\beta = \delta\tau\mathrm{N}$, where δ is the number of contacts that each infected has with susceptible individuals, τ is the fraction of contacts that result in a transmission (transmissibility), and N is the total population size. The removal rates for symptomatic and asymptomatic individuals are $\gamma > 0$ and $\nu > 0$, respectively. In other words, the average number of days between infection and recovery/death of symptomatic individuals is γ^{-1} and of asymptomatic individuals is ν^{-1}. On the other hand, $\rho > 0$ represents the fraction of symptomatic infected individuals. These parameters do not depend on the population size, just on the characteristics of the disease and the social behavior.

Since it is assumed that the population is constant, we can express:

$$S(t) + I(t) + A(t) + R(t) = \mathrm{N}, \quad \forall t \geq 0 \tag{3.65}$$

However, to simplify, let us consider normalized populations, i.e., $s(t) = S(t)/\mathrm{N}$, $i(t) = I(t)/\mathrm{N}$, $a(t) = A(t)/\mathrm{N}$ and $r(t) = R(t)/\mathrm{N}$. With the normalized variables, the model (3.64) is now

$$\begin{aligned}
\dot{s}(t) &= -\beta s(t)\left[a(t) + i(t)\right], \\
\dot{i}(t) &= \rho\beta s(t)\left[a(t) + i(t)\right] - \gamma i(t), \\
\dot{a}(t) &= (1-\rho)\,\beta s(t)\left[a(t) + i(t)\right] - \nu a(t), \\
\dot{r}(t) &= \gamma i(t) + \nu a(t), \\
y(t) &= i(t),
\end{aligned} \tag{3.66}$$

with initial conditions: $s(0) = s_0$, $i(0) = i_0$, $a(0) = a_0$ and $r(0) = r_0$. Meanwhile, (3.65) is:

$$s(t) + i(t) + a(t) + r(t) = 1, \quad \forall t \geq 0 \tag{3.67}$$

The basic reproduction number $\mathscr{R}_0 = \delta\tau/(\gamma + \nu)$ and the effective reproductive number $\mathscr{R}_e = s(0)\delta\tau/(\gamma + \nu)$ are two helpful parameters to describe any disease spread. $\mathscr{R}_0$ is an estimate of how many new infections are originated from a single

[9] The A-SIR model assumes permanent immunity of individuals who recovered and a constant population (no natural births or deaths). On the other hand, the removed population of this model includes any individual that was infected and has recovered or died.

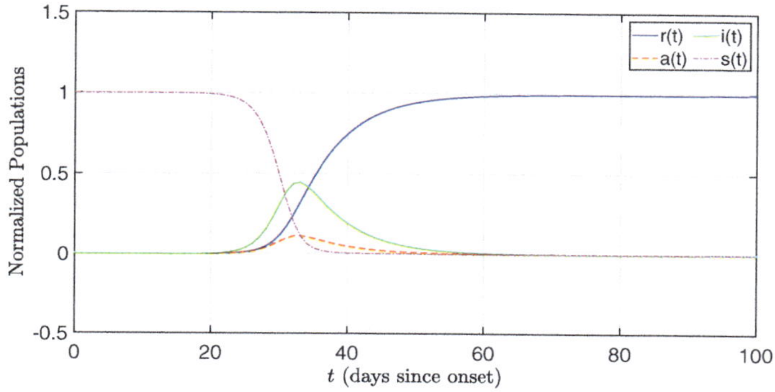

(a)

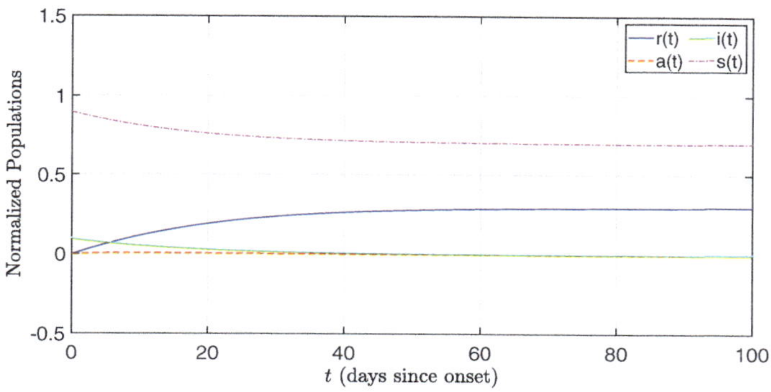

(b)

Fig. 3.4 A-SIR model: normalized population's evolution with $\mathscr{R}_e > 1$ (**a**) and with $\mathscr{R}_e \leq 1$ (**b**). In **a**, we can observe an epidemic scenario; meanwhile in **b**, the infected population decreases monotonically to zero. Both cases reach an endemic equilibrium, each one with different characteristics

infected individual in the initial phase of the pandemic. $\mathscr{R}_e$ is the average number of secondary cases per infected individual.[10]

Theorem 3.2 *[40] If $\mathscr{R}_e \leq 1$, then the infected population decreases monotonically to zero as $t \to \infty$. If $\mathscr{R}_e > 1$, then the infected population starts increasing, reaches its maximum and then decreases to zero as $t \to \infty$.*[11]

Observe in Fig. 3.4 the evolution of the A-SIR model. One can notice the influence of the effective reproduction number. In Fig. 3.4a, the total population is initially susceptible, and therefore, $\mathscr{R}_e \approx \mathscr{R}_0 > 1$. This is the case for the COVID-19 pandemic,

[10] If the entire population is initially susceptible, i.e., $s(0) \approx 1$, then $\mathscr{R}_e$ is approximately equal to $\mathscr{R}_0$. This assumption is reasonable for a new virus such as the SARS-CoV-2, where a vaccine was not initially available and information about innate immunity is still unclear [19].

[11] Any disease can cause an epidemic if $\mathscr{R}_0 > 1$ or $\delta\tau > \gamma + \nu$. Moreover, both cases reach an endemic equilibrium point, i.e., when $\dot{s} = \dot{i} = \dot{a} = \dot{r} = 0$.

where researchers have reported values between 2 and 6 for the basic reproduction number [1, 23, 31]. On the other hand, in Fig. 3.4b it observed the case when $\mathscr{R}_e \leq 1$. An example of this scenario is the Nipah virus, whose basic reproduction number is 0.48 [26].

3.3.2 Observer Construction

Let us define the auxiliary variable $\eta(x)$ for the A-SIR model as

$$\eta(x) = \beta s(t)\,[a(t) + i(t)] \tag{3.68}$$

Therefore, we can express the model (3.66) as

$$\begin{aligned}
\dot{s}(t) &= -\,\eta(x),\\
\dot{i}(t) &= \rho\eta(x) - \gamma i(t),\\
\dot{a}(t) &= (1-\rho)\,\eta(x) - \nu a(t),\\
\dot{\eta}(x) &= \Omega(x),\\
\dot{r}(t) &= \gamma i(t) + \nu a(t),\\
y(t) &= i(t),
\end{aligned} \tag{3.69}$$

where $x(t) = [s, i, a, \eta, r]^T$ and $\Omega(x)$ is the unknown dynamics of $\eta(x)$. Recall that the PI observer requires some assumptions to be satisfied. In particular, we have that the unknown dynamics is bounded for obvious reasons. On the other hand, we have that our auxiliary variable is algebraically observable since

$$\eta = \frac{1}{\rho}\left(\dot{y} + \gamma y\right), \quad \rho \neq 0 \tag{3.70}$$

For comparison purposes, let us discard the integral part of PI observer, such that the following cases are presented.

Case 3.1 The Proportional observer is

$$\dot{\hat{\eta}} = k\left(\eta - \hat{\eta}\right), \tag{3.71}$$

where for simplicity, $k_p = k$. Now, substituting (3.70) in (3.71), it is obtained

$$\dot{\hat{\eta}} = k\left[\frac{1}{\rho}\left(\dot{y} + \gamma y\right) - \hat{\eta}\right] \tag{3.72}$$

Then, to avoid the derivative of y, let us propose an artificial variable, denoted by α, such that

$$\alpha = \hat{\eta} - \frac{k}{\rho} y, \tag{3.73}$$

Thus, with the derivative of the artificial variable and expression (3.72),

$$\dot{\alpha} = \frac{k\gamma}{\rho} y - k\hat{\eta} \tag{3.74}$$

Since $\hat{\eta} = \alpha + \frac{k}{\rho} y$, the Proportional reduced-order state observer for unknown dynamics of system (3.69) is given by

$$\begin{aligned} \dot{\alpha} &= -k\alpha + \frac{y}{\rho}\left(k\gamma - k^2\right), \quad \alpha(0) = \alpha_0, \\ \hat{\eta} &= \alpha + \frac{k}{\rho} y \end{aligned} \tag{3.75}$$

and finally, the unknown populations are

$$\begin{aligned} \hat{s}(t) &= -\int_0^t \hat{\eta}(x(\tau)) d\tau, \\ \hat{a}(t) &= \frac{\hat{\eta}(x)}{\beta \hat{s}(t)} - y(t), \quad \hat{s}(t) \neq 0 \\ \hat{r}(t) &= 1 - y(t) - \hat{a}(t) - \hat{s}(t) \end{aligned}$$

Remark 3.2 Notice that for this problem the population estimation errors are defined as $\tilde{s}(t) = s(t) - \hat{s}(t)$, $\tilde{a}(t) = a(t) - \hat{a}(t)$ and $\tilde{r}(t) = r(t) - \hat{r}(t)$.

Case 3.2 For this particular problem, the model-free reduced-order observer with integral and proportional actions can be expressed as

$$\begin{aligned} \dot{\hat{\eta}} &= k_p \left[\frac{1}{\rho}(\dot{y} + \gamma y) - \hat{\eta}\right] + k_i \hat{\eta}_1, \\ \dot{\hat{\eta}}_1 &= \frac{1}{\rho}(\dot{y} + \gamma y) - \hat{\eta} \end{aligned} \tag{3.76}$$

where the artificial variables are

$$\begin{aligned} \alpha_1 &= \hat{\eta} - \frac{k_p}{\rho} y, \\ \alpha_2 &= \hat{\eta}_1 - \frac{k_i}{\rho} y, \end{aligned} \tag{3.77}$$

such that it is obtained

$$
\begin{aligned}
\dot{\alpha}_1 =& \frac{k_p}{\rho}\gamma y - k_p\hat{\eta} + \hat{\eta}_1, \\
\dot{\alpha}_2 =& \frac{k_i}{\rho}\gamma y - k_i\hat{\eta}
\end{aligned}
\tag{3.78}
$$

Now, by considering (3.77), and after some algebraic manipulations, (3.78) is expressed as

$$
\begin{aligned}
\dot{\alpha}_1 &= -k_p\alpha_1 + \alpha_2 + \frac{y}{\rho}\left(k_i + k_p\gamma - k_p^2\right), \\
\dot{\alpha}_2 &= -k_i\alpha_1 + \frac{k_i}{\rho}y\left(\gamma - k_p\right),
\end{aligned}
\tag{3.79}
$$

such that the model-free PI reduced-order state observer is

$$
\begin{aligned}
\dot{\alpha}_1 &= -k_p\alpha_1 + \alpha_2 + \frac{y}{\rho}\left(k_i + k_p\gamma - k_p^2\right), \quad \alpha_{1_0} = \alpha_1(0), \\
\dot{\alpha}_2 &= -k_i\alpha_1 + \frac{k_i}{\rho}y\left(\gamma - k_p\right), \quad \alpha_{2_0} = \alpha_2(0), \\
\hat{\eta} &= \alpha_1 + \frac{k_p}{\rho}y, \\
\hat{\eta}_1 &= \alpha_2 + \frac{k_i}{\rho}y
\end{aligned}
\tag{3.80}
$$

Thus, the unknown populations are given by

$$
\begin{aligned}
\hat{s}(t) &= -\int_0^t \hat{\eta}(x(\tau))d\tau, \\
\hat{a}(t) &= \frac{\hat{\eta}(x)}{\beta\hat{s}(t)} - y, \quad \hat{s}(t) \neq 0 \\
\hat{r}(t) &= 1 - y(t) - \hat{a}(t) - \hat{s}(t)
\end{aligned}
$$

3.3.3 Estimation with Environmental Noise

Evidence suggests that the COVID-19 pandemic is highly influenced by external conditions such as humidity or temperature. It has been reported that cold and dry conditions increase the transmission rate and, therefore, boost the virus spread [18, 36]. In addition, the effect of climatic fluctuations in the A-SIR model results in poor measurements. To model this, let us consider the following expression [44]

$$
\omega g(t) = \omega\beta s(t)\left[a(t) + i(t)\right]^{1/2}, \tag{3.81}
$$

where ω is a random number that belongs to the normal distribution, i.e., $\omega \in (0, 1)$. On the other hand, $g(t)$ is an analytic function. Let us consider that the system's output is affected by this additive signal, such that we have

$$y(t) = i(t) + \omega g(t), \tag{3.82}$$

One can notice that $\omega g(t)$ is a nonevanescent term and increases as the number of infected individuals increases. In addition, let us assume that the additive term is bounded, i.e., $\|\omega g(t)\| \leq M$.

It is possible to prove that the estimation error of the PI reduced-order observer is stable despite the environmental noise. Thus, the following theorem is established.

Theorem 3.3 *The estimation error of the model-free Proportional-Integral reduced-order observer, for the unknown dynamics of system* (3.69), *with output* (3.82), *is ultimate uniformly bounded.*

Proof Let us express the output (3.82) as $y(t) = y_1(t) + wg(t)$. Moreover, the auxiliary variable η can be selected as a sum of two differential polynomials, i.e.,

$$\eta = P_1(y_1, \dot{y}_1, \ldots, y_1^{(\mu_1)}) + \omega P_2(g, \dot{g}, \ldots, g^{(\mu_2)}), \quad \mu_1, \mu_2 \in \mathbb{N}, \tag{3.83}$$

where $P_1 \in C^1$ and $P_2 \in C^1$. Thus, the time derivative of η can be expressed as

$$\dot{\eta} = \Omega(x) + \omega\Psi(t) \tag{3.84}$$

Then, since the estimation error is $e = \eta - \hat{\eta}$, the dynamics of the estimation error is

$$\begin{aligned} \dot{e} &= \Omega(x) - k_p e - k_i \hat{\eta}_1 + \omega\Psi(t), \\ \dot{\hat{\eta}}_1 &= e \end{aligned} \tag{3.85}$$

or in matrix form

$$\dot{E} = \mathrm{K}E + \bar{\Omega} + \omega\bar{\Psi}, \tag{3.86}$$

where

$$\mathrm{K} = \begin{pmatrix} -k_p & -k_i \\ 1 & 0 \end{pmatrix}, E = \begin{pmatrix} e \\ \hat{\eta}_1 \end{pmatrix}, \bar{\Omega} = \begin{pmatrix} \Omega \\ 0 \end{pmatrix}, \bar{\Psi} = \begin{pmatrix} \Psi \\ 0 \end{pmatrix}$$

Let us consider the following Lyapunov candidate function

$$V(E) = E^T P E + \int_0^t \|\bar{\Omega}(x(\tau))\|^2 d\tau \tag{3.87}$$

where $P = P^T > 0$. The derivative of the Lyapunov function is given by

$$\begin{aligned}\dot{V}(E) &= \dot{E}^T P E + E^T P \dot{E} + \|\Omega\|^2 \\ &= \left[E^T \mathrm{K}^T + \bar{\Omega}^T + \omega\bar{\Psi}^T\right] P E + E^T P \left[\mathrm{K}E + \bar{\Omega} + \omega\bar{\Psi}\right] + \bar{\Omega}^T \bar{\Omega} \\ &= E^T \mathrm{K}^T P E + E^T P \mathrm{K} E + \bar{\Omega}^T P E + E^T P \bar{\Omega} \\ &\quad + \bar{\Omega}^T \bar{\Omega} + \omega\bar{\Psi}^T P E + \omega E^T P \bar{\Psi},\end{aligned} \tag{3.88}$$

which can be expressed as

$$\dot{V}(E) = \bar{E}^T Q \bar{E} + 2\omega\Phi^T R \bar{E}, \tag{3.89}$$

where

$$\bar{E} = \begin{pmatrix} E \\ \bar{\Omega} \end{pmatrix}, \; \Phi^T = \begin{pmatrix} \Psi \\ 0 \\ 0 \\ 0 \end{pmatrix}, \; R = \begin{pmatrix} P & 0_2 \\ 0_2 & 0_2 \end{pmatrix},$$

$$Q = \begin{pmatrix} K^T P + P K & P \\ P & I \end{pmatrix} < 0$$

Then, from the Rayleigh inequality,

$$\lambda_{\min}(Q)\|\bar{E}\|^2 \leq \bar{E}^T Q \bar{E} \leq \lambda_{\max}(Q)\|\bar{E}\|^2 \tag{3.90}$$

Since the additive noise is bounded, $\|\omega\Phi\| \leq M$. Therefore,

$$\|2\omega\Phi^T R \bar{E}\| \leq 2M\|\bar{E}\|_R = 2M(\lambda_{\max}(R))^{1/2}\|\bar{E}\| \tag{3.91}$$

such that

$$\dot{V}(E) \leq \lambda_{\max}(Q)\|\bar{E}\|^2 + 2M(\lambda_{\max}(R))^{1/2}\|\bar{E}\| \tag{3.92}$$

Thus, we can apply the uniform ultimate boundedness theorem [10]. That is, $\bar{E}(t)$ is uniformly bounded for any initial condition $\bar{E}(0)$, and even more, $\bar{E}(t)$ remains in the compact set $B_{\bar{\delta}} = \left\{\bar{E} : \|\bar{E}\| \leq \bar{\delta}, \bar{\delta} > 0\right\}$, where

$$\bar{\delta} = \left(\frac{\lambda_{\max}(R)}{\lambda_{\min}(R)}\right)^{1/2} \left(\frac{2M\,(\lambda_{\max}(R))^{1/2}}{\lambda_{\max}(Q)}\right) > 0. \tag{3.93}$$

Hence, the estimation error of the Proportional-Integral reduced-order observer is uniform, ultimate bounded in the presence of additive noise. □

3.4 Pandemic Application: Simulation Results

The performance of the proposed observer can be shown by considering three cases: The first case considers as output (and therefore as available data) the numerical solution of the A-SIR model. The second case also considers the numerical solution, however, it assumed the presence of additive noise in the output (environmental noise). Meanwhile, in the third case the available data is considered to be real data reported by authorities. It is worth mentioning that all cases are focused in the analysis of the COVID-19 pandemic in Mexico City.

3.4.1 Estimation with Numerical Solution

Consider the numerical solution of the A-SIR model with no control (intervention) action, i.e., the natural progression of the COVID-19 pandemic. The A-SIR model parameters used are shown in Table 3.1. On the other hand, according to the last census realized by the Instituto Nacional de Estadística y Geografía (INEGI), Mexico City, total population is $\mathrm{N} = 9,209,944$ [20].

The parameters β, γ and ν are statistical estimates made by considering only official data from Mexico City. The parameter ρ was inferred from a serological study conducted in New York City. The simulation of the A-SIR model was realized with the following initial conditions: $i(0) = 1/\mathrm{N}$, $a(0) = 1/\mathrm{N}$, $r(0) = 0$ and $s(0) = 1 - i(0) - a(0)$. In Fig. 3.5, the numerical solution of all the normalized populations of the A-SIR model is observed. Then, the numerical solution $i(t)$ is the available data for each observer.

Table 3.2 contains the initial conditions used for the different observers, as well as the gains for each. We use the same parameters from Table 3.1. As a point of reference, we include the estimations made by the well-known Luenberger observer [27]. This observer is extensively used due to its simplicity. However, this observer is not model-free; therefore we need to proportionate initial conditions for $\hat{s}$, $\hat{i}$, $\hat{a}$ and $\hat{r}$. The parameter L is known as the Luenberger gain.

Table 3.1 A-SIR model simulation parameters

Parameter description	Symbol	Value
Disease transmission rate [33]	β	0.46
Fraction of infected symptomatic individuals [35]	ρ	0.15
Symptomatic removal rate (days^{-1}) [33]	γ	1/4.86
Asymptomatic removal rate (days^{-1}) [33]	ν	1/4.86

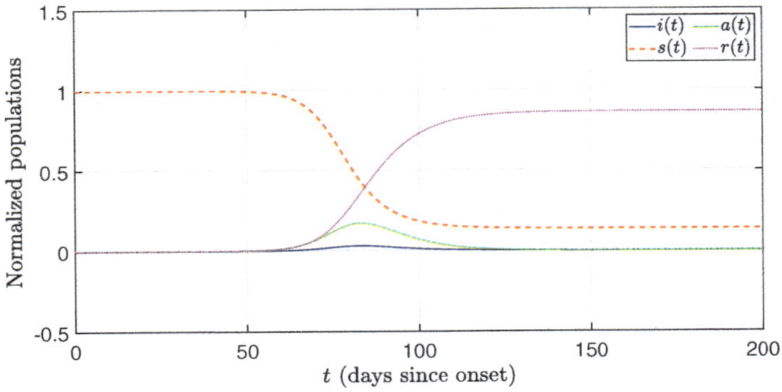

Fig. 3.5 A-SIR model numerical solution: natural progression of the COVID-19 pandemic. Observe that in the endemic equilibrium point remains a considerable amount of susceptible individuals. On the other hand, infected individuals reach a maximum and then converge to zero. This behavior corresponds to a pandemic with $\mathscr{R}_0 > 1$, just as is the case of the SARS-CoV-2 virus

Table 3.2 Initial conditions and gains of the different state observers

Proportional	Proportional-Integral	Luenberger
$\alpha(0) = 0.1$	$\alpha_1(0) = 1$	$\hat{i}(0) = 1/N$
$\eta(0) = 1$	$\alpha_2(0) = 0$	$\hat{a}(0) = 1/N$
$k = 25$	$\eta(0) = 1$	$\hat{r}(0) = 0$
	$k_p = 25$	$\hat{s}(0) = 1 - i(0) - a(0)$
	$k_i = 20$	$L = 250$

In Fig. 3.6, the estimations made by the different observers are shown. In general, we notice that all the observers are capable to estimate the normalized population of asymptomatic, removed and susceptible individuals. Nonetheless, Fig. 3.7 reveals a considerable discrepancy between the estimate made by Luenberger observer and the numerical solution of the A-SIR model, especially in the susceptible population case. Besides, an overshoot is contained at the beginning of the estimates made by the Proportional and Proportional-Integral estimations algorithms. Due to this, the estimates take negative values for a time. However, these quickly converge to numerical solutions. These overshoots are a result of the initial conditions. As the initial conditions get closer to the real initial conditions, the overshoots decrease.

Figure 3.8 displays the behavior of each auxiliary variable(s) η and artificial variable(s) α for Proportional and Proportional-Integral observers. Remember that from these variables the populations or variables of interest are reconstructed.

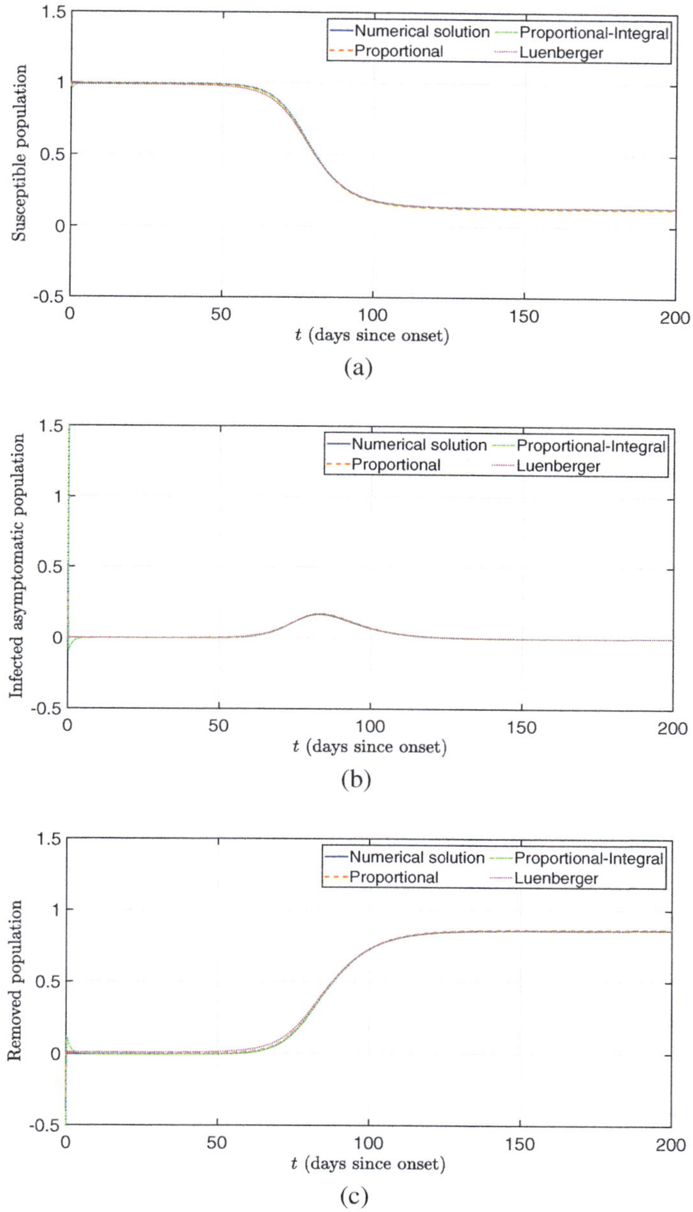

Fig. 3.6 Population estimates by considering the numerical solution of the A-SIR model

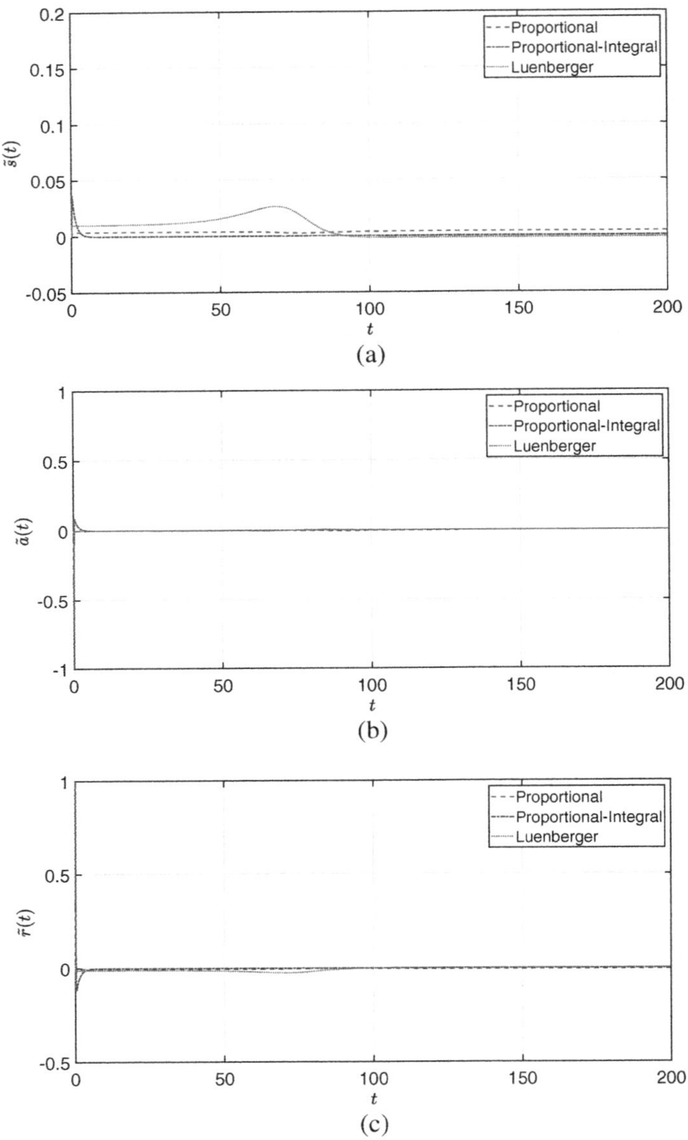

Fig. 3.7 Population estimation errors by considering the numerical solution of the A-SIR model

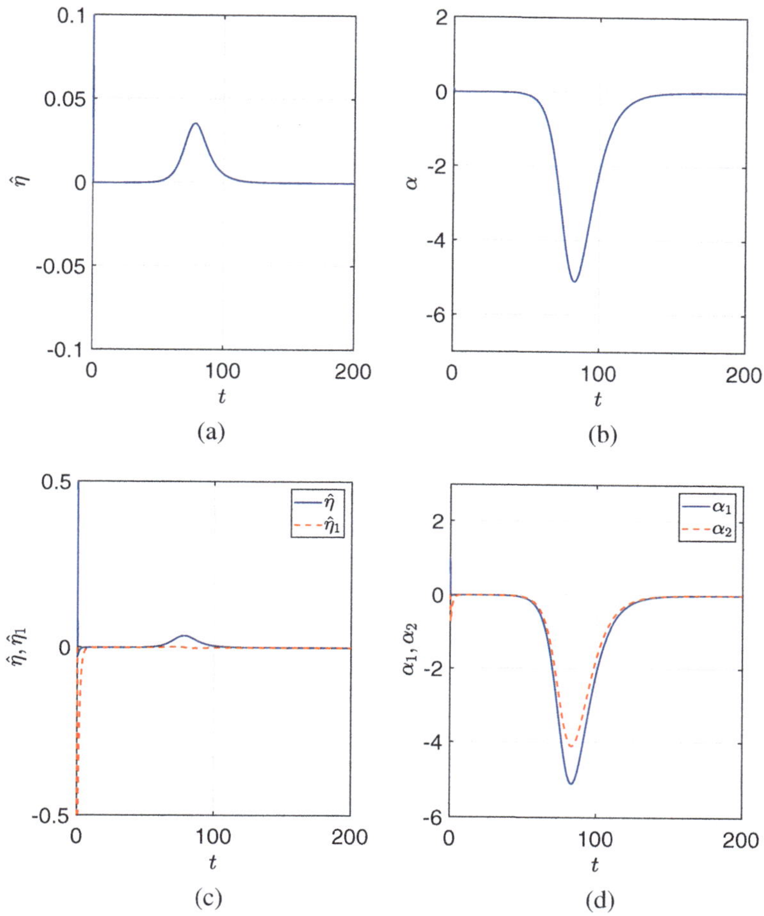

Fig. 3.8 Auxiliary variable(s) η and artificial variable(s) α. **a**, **b** Proportional observer and **c**, **d** Proportional-Integral observer

3.4.2 Numerical Solution and Environmental Noise

Now, let us consider that the output of the A-SIR model contains an additive noise term $\omega g(t)$. Function $g(t)$ is as in (3.81). In this scenario, the Luenberger observer is omitted, since this is not appropriate to deal with additive noise. The simulation parameters and initial conditions are the same from Tables 3.1 and 3.2, respectively. The only change is the values of the gains, which are set: $k = 50$, $k_p = 1.4$ and $k_i = 0.8$.

In Figs. 3.9 and 3.10, it is shown that the estimates of both algorithms differ considerably from the A-SIR model numerical solution when the number of infected

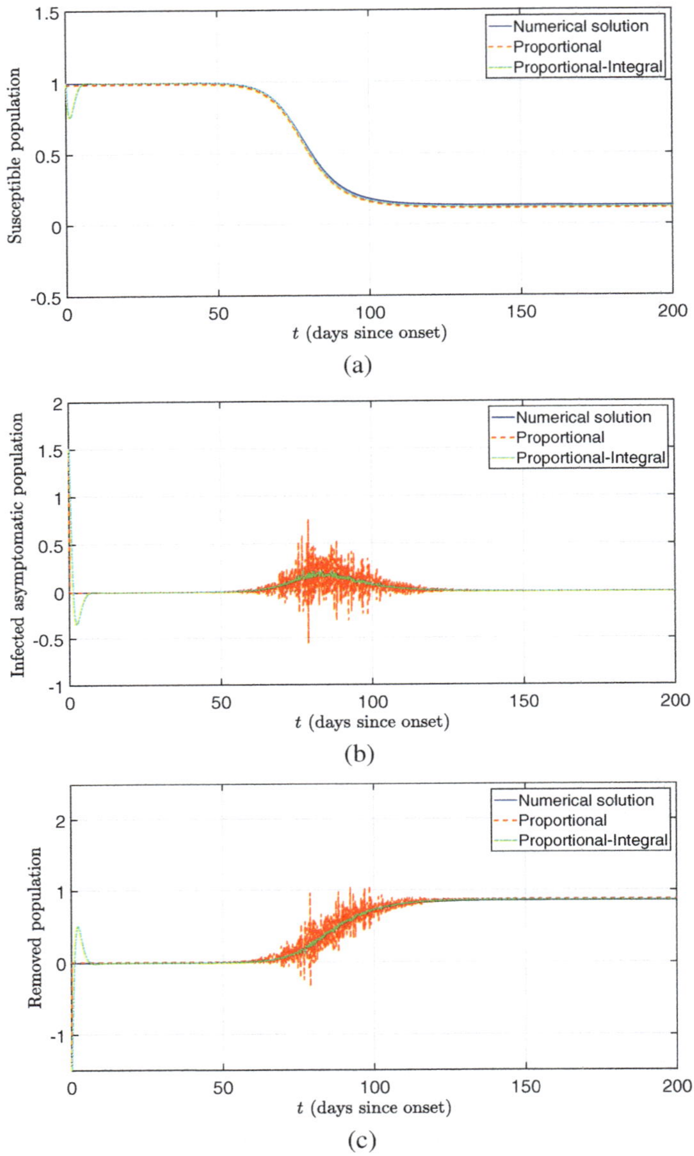

Fig. 3.9 Population estimates with additive environmental noise

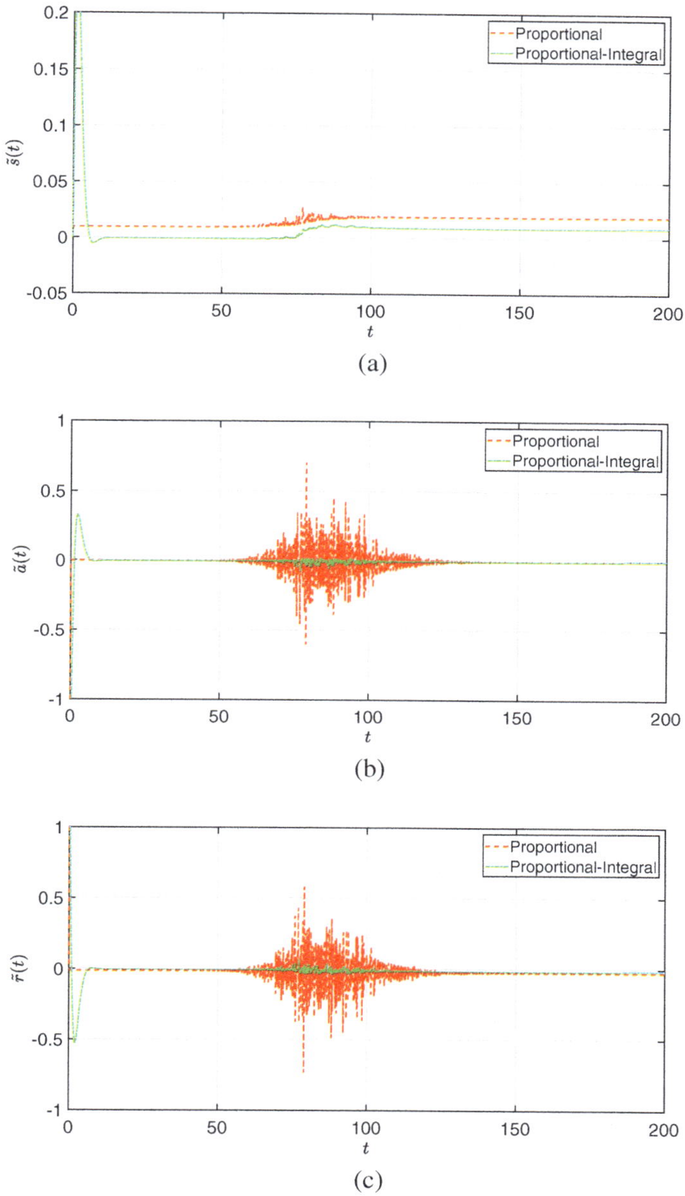

Fig. 3.10 Population estimation errors with additive environmental noise

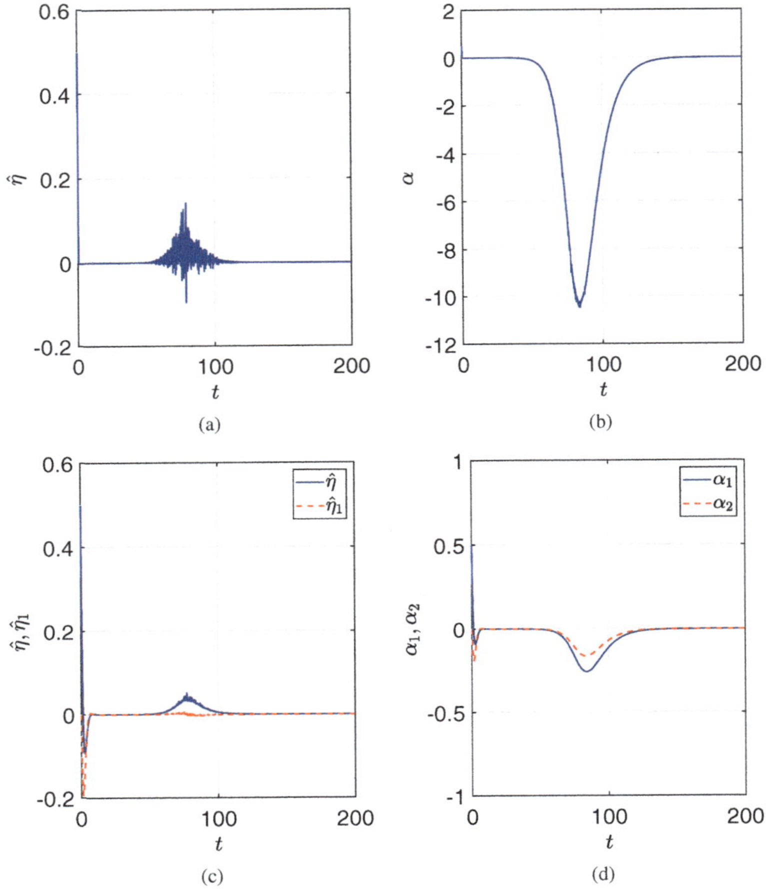

Fig. 3.11 Auxiliary variable(s) η and artificial variable(s) α with additive environmental noise. **a, b** Proportional observer and **c, d** Proportional-Integral observer

cases is maximum. This is due to the increment of the noise term. On the other hand, although the Proportional-Integral observer presents overshooting, its estimates are better than those made by the Proportional observer. This is due to the attenuation effect given by the integral action. In Fig. 3.11, this effect can be appreciated better. Notice how the artificial variable α from the Proportional estimation algorithm is clearly under the effect of the noise term. On the other hand, the artificial variables from the Proportional-Integral observer almost lack this noise.

The population estimation errors in Fig. 3.10 can be observed. As was established in Theorem 3.3, these errors are ultimate uniformly bounded, although the population estimation errors of the Proportional observer do not exhibit overshooting.

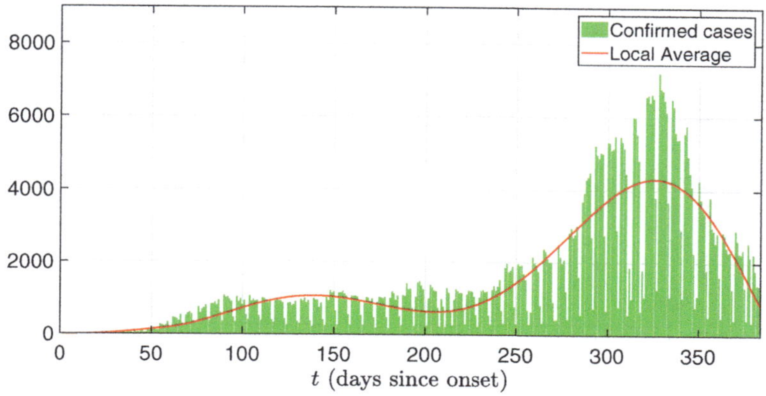

Fig. 3.12 COVID-19 pandemic in Mexico City: infected cases reported from February 22, 2020, to March 13, 2021

3.4.3 Estimation with Reported Data

Now, official data of infected individuals reported in Mexico City from February 22, 2020, to March 13, 2021, is considered [15]. Since the testing rate in Mexico is low (0.29 tests per thousand people, maximum value registered [30]), it is assumed that all reported cases are symptomatic. In Fig. 3.12, the infected cases reported in Mexico City are displayed. Mexico City has carried out two confinements during the pandemic; as a result, the behavior of the symptomatic population is quite different from that shown in Fig. 3.5. One can notice that the reported data is noisy. However, through a moving/local average, a smooth curve can be obtained. Both curves can be used to make population estimations.

The Proportional-Integral observer is used to make the estimates with the real data since this showed the best performance. For the observer, the initial conditions from Table 3.2 and the gains $k_p = 100$ and $k_i = 50$ are used.

Figure 3.13 shows the estimates obtained by the algorithm. Discontinuous red lines are the estimates when we fed the observer with the original data. On the other hand, the continuous blue lines are the estimates with the smooth data. Both cases exhibit overshooting at the beginning; however, these seem to converge to a common trajectory. The local average improves greatly the smoothness of the estimates, especially the estimation of the asymptotic population (Fig. 3.13b).

In this case, there is no information about the evolution of susceptible, removed or asymptomatic populations during the pandemic. Therefore, it is not possible to compare directly the obtained estimates. However, there is additional data that might be useful to tell how precise is the estimate of removed individuals.

In Fig. 3.13c, the accumulated deceased cases in Mexico City are shown (discontinuous black line). Until March 13, the number of deceased people in Mexico City was $29,047$. On the other hand, the accumulated reported cases until the

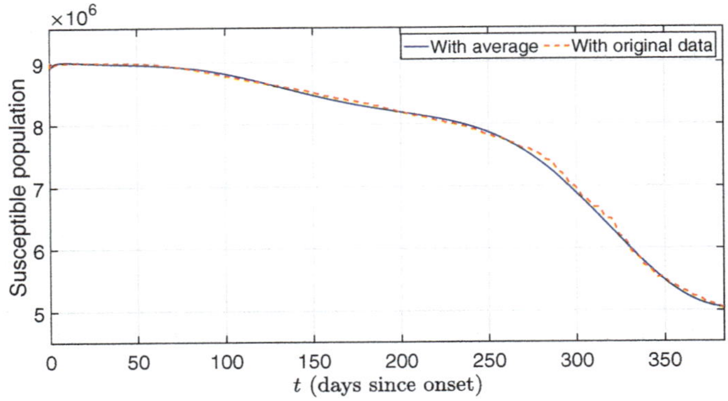

(a)

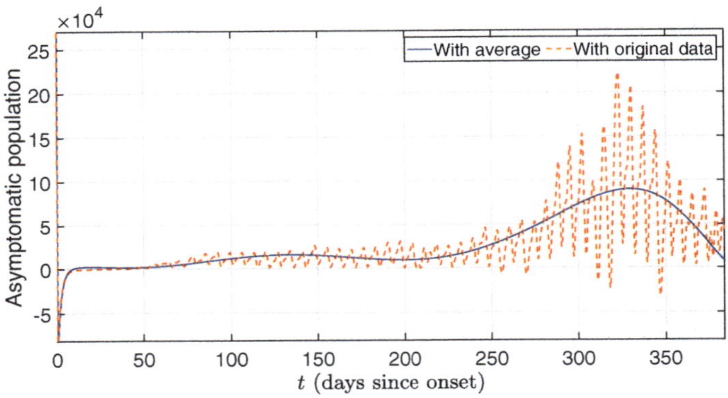

(b)

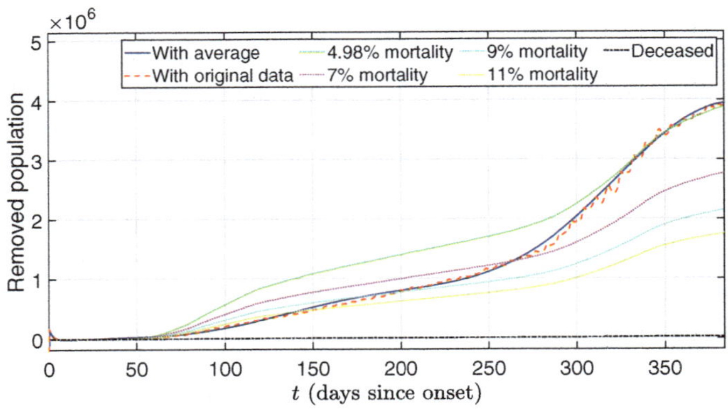

(c)

Fig. 3.13 Proportional-Integral estimates with real infected cases reported in Mexico City

same date was 583, 698. [15]. Meanwhile, for the same date, the reported mortality rate (ratio between confirmed deaths and confirmed cases) was 4.98% [22]. Moreover, if we consider that the fraction of infected individuals that are symptomatic is 0.15, then we can multiply the number of accumulated deceased cases by the factor $f = (100/M_r) * (100/15)$, where M_r is the mortality rate.

The projected curves on Fig. 3.13c correspond to dotted lines. Notice that the estimates of removed individuals and the projected curve with a mortality rate of 4.98% are similar for the last 50 days. The disparity before those days might be a consequence of the mortality rate, which is continuously changing. For example, the maximum value registered for Mexico is 12.4% (June 2020) and in March 2021 is 9%.

The dotted green line indicates that the projection of removed individuals for March 13, 2021, is 3,888,350. Meanwhile, for the same day, the estimate number of removed individuals is 3,954,610 (with local average data) and 3,934,790 (with original data). Then, on day 384, the relative error for both cases is about 0.002. According to our estimates, until March 13, 2021, around 55% of the population in Mexico City is still susceptible to infection.

Note that all these estimates depend on the parameters to be considered. In particular, the fraction of symptomatic individuals is based on a study carried out in New York City, so this value could be different in Mexico City. As can be seen with this application case, the PI observer proposed is capable of dealing with real-world problems and is not limited to chaotic systems.

References

1. Abbott, S., Hellewell, J., Munday, J., Chun, J., Thompson, R., Bosse, N., Chan, Y., Russell, T., Jarvis, C., nCov Working Group, C., et al.: Temporal variation in transmission during the Covid-19 outbreak. In: CMMID Repository (2020)
2. Acemoglu, D., Chernozhukov, V., Werning, I., Whinston, M.D.: A multi-risk sir model with optimally targeted lockdown. NBER Working Paper (2020)
3. Bai, Y., Yao, L., Wei, T., Tian, F., Jin, D.-Y., Chen, L., Wang, M.: Presumed asymptomatic carrier transmission of covid-19. JAMA **323**(14), 1406–1407 (2020)
4. Bambra, C., Riordan, R., Ford, J., Matthews, F.: The covid-19 pandemic and health inequalities. J. Epidemiol. Commun. Health **74**(11), 964–968 (2020)
5. Bilotta, E., Pantano, P., Stranges, F.: A gallery of Chua attractors: part I. Int. J. Bifurcat. Chaos **17**(01), 1–60 (2007)
6. Brassey, J., Heneghan, C., Mahtani, K., Aroson, J.: Do Weather Conditions Influence the Transmission of the Coronavirus (SARS-COV-2)? (2020)
7. Chen, D., Yang, Y., Zhang, Y., Yu, W.: Prediction of covid-19 spread by sliding MSEIR observer. Sci. China Inf. Sci. **63**(12), 1–13 (2020)
8. Claude, D., Fliess, M., Isidori, A.: Immersion directe et par bouclage d'un système non linéaire dans un linéaire. CR Acad. Sci. Paris **296**(1), 237–240 (1983)
9. Cooper, I., Mondal, A., Antonopoulos, C.G.: A sir model assumption for the spread of Covid-19 in different communities. Chaos, Solitons Fractals **139**, 110057 (2020)
10. Corless, M., Leitmann, G.: Continuous state feedback guaranteeing uniform ultimate boundedness for uncertain dynamic systems. IEEE Trans. Autom. Control **26**(5), 1139–1144 (1981)

11. Díaz, J., Antonio-López-Bueno, J., Culqui, D., Asensio, C., Sánchez-Martínez, G., Linares, C.: Does exposure to noise pollution influence the incidence and severity of Covid-19? Environ. Res. **195**, 110766 (2021)
12. Fauci, A.S., Lane, H.C., Redfield, R.R.: Covid-19-Navigating the Uncharted (2020)
13. Furukawa, N.W., Brooks, J.T., Sobel, J.: Evidence supporting transmission of severe acute respiratory syndrome coronavirus 2 while presymptomatic or asymptomatic. Emerg. Infect. Dis. **26**, 7 (2020)
14. Gaeta, G.: A simple sir model with a large set of asymptomatic infectives. arXiv preprint arXiv:2003.08720 (2020)
15. Gobierno de México, and Consejo Nacional de Ciencia y Tecnología: Covid-19 México (2021)
16. He, S., Peng, Y., Sun, K.: Seir modeling of the Covid-19 and its dynamics. Nonlinear Dyn. **101**(3), 1667–1680 (2020)
17. Heneghan, C., Jefferson, T.: Effect of Latitude on Covid-19 (2020)
18. Holtmann, M., Jones, M., Shah, A., Holtmann, G.: Low ambient temperatures are associated with more rapid spread of Covid-19 in the early phase of the endemic. Environ. Res. (2020)
19. Hosseini, A., Hashemi, V., Shomali, N., Asghari, F., Gharibi, T., Akbari, M., Gholizadeh, S., Jafari, A.: Innate and adaptive immune responses against coronavirus. Biomed. Pharmacother. 110859 (2020)
20. Instituto Nacional de Estadística y Geografía: México en cifras: Ciudad de México (2021)
21. Ivorra, B., Ferrández, M., Vela-Pérez, M., Ramos, A.: Mathematical modeling of the spread of the coronavirus disease 2019 (Covid-19) considering its particular characteristics. The case of China. Commun. Nonlinear Sci. Numer. Simul. **88**, 105303 (2020)
22. Johns Hopkins University: Covid-19 Dashboard by the Center for Systems Science and Engineering (CSSE) at Johns Hopkins University (2021)
23. Katul, G.G., Mrad, A., Bonetti, S., Manoli, G., Parolari, A.J.: Global convergence of Covid-19 basic reproduction number and estimation from early-time sir dynamics. PLoS ONE **15**(9), e0239800 (2020)
24. Kitsos, C., Besancon, G., Prieur, C.: High-gain observer design for a class of quasi-linear integro-differential hyperbolic system-application to an epidemic model. IEEE Trans. Autom. Control **2021** (2021)
25. Liu, Y., Gayle, A.A., Wilder-Smith, A., Rocklöv, J.: The reproductive number of Covid-19 is higher compared to SARS coronavirus. J. Travel Med. (2020)
26. Luby, S.P., Hossain, M.J., Gurley, E.S., Ahmed, B.-N., Banu, S., Khan, S.U., Homaira, N., Rota, P.A., Rollin, P.E., Comer, J.A., et al.: Recurrent zoonotic transmission of Nipah virus into humans, Bangladesh, 2001–2007. Emerg. Infect. Dis. **15**(8), 1229 (2009)
27. Luenberger, D.G.: Observing the state of a linear system. IEEE Trans. Mil. Electron. **8**(2), 74–80 (1964)
28. Monteiro, L., Fanti, V., Tessaro, A.: On the spread of Sars-Cov-2 under quarantine: a study based on probabilistic cellular automaton. Ecol. Complex. **44**, 100879 (2020)
29. Nijmeijer, H., Mareels, I.M.: An observer looks at synchronization. IEEE Trans. Circ. Syst. I: Fundam. Theory Appl. **44**(10), 882–890 (1997)
30. Oxford University, Global Change Data Lab: Coronavirus Pandemic Country Profile. Mexico (2021)
31. Park, M., Cook, A.R., Lim, J.T., Sun, Y., Dickens, B.L.: A systematic review of Covid-19 epidemiology based on current evidence. J. Clin. Med. **9**(4), 967 (2020)
32. Péni, T., Csutak, B., Szederkényi, G., Röst, G.: Nonlinear model predictive control with logic constraints for Covid-19 management. Nonlinear Dyn. **102**(4), 1965–1986 (2020)
33. Prieto, K., Chavez-Hernandez, M., Romero-Leiton, J.P.: On mobility trends analysis of Covid-19 dissemination in Mexico city. medRxiv (2021)
34. Read, J.M., Bridgen, J.R., Cummings, D.A., Ho, A., Jewell, C.P.: Novel coronavirus 2019-ncov: early estimation of epidemiological parameters and epidemic predictions. MedRxiv (2020)
35. Subramanian, R., He, Q., Pascual, M.: Quantifying asymptomatic infection and transmission of Covid-19 in New York city using observed cases, serology, and testing capacity. Proc. Natl. Acad. Sci. **118**, 9 (2021)

36. Tobías, A., Molina, T.: Is temperature reducing the transmission of Covid-19? Environ. Res. **186**, 109553 (2020)
37. Toda, A.A.: Susceptible-infected-recovered (sir) dynamics of Covid-19 and economic impact. arXiv preprint. arXiv:2003.11221 (2020)
38. Wang, Y., Wang, Y., Chen, Y., Qin, Q.: Unique epidemiological and clinical features of the emerging 2019 novel coronavirus pneumonia (Covid-19) implicate special control measures. J. Med. Virol. **92**(6), 568–576 (2020)
39. Wei, W., Duan, B., Zuo, M., Zhu, Q.: An extended state observer based u-model control of the Covid-19. ISA Trans. (2021)
40. Weiss, H.H.: The sir model and the foundations of public health. Mater. Math. 0001–17 (2013)
41. World Health Organization: What Do We Know About Sars-Cov-2 and Covid-19? (2020)
42. Yao, Y., Pan, J., Wang, W., Liu, Z., Kan, H., Qiu, Y., Meng, X., Wang, W.: Association of particulate matter pollution and case fatality rate of Covid-19 in 49 Chinese cities. Sci. Total Environ. **741**, 140396 (2020)
43. Zhang, J., Litvinova, M., Liang, Y., Wang, Y., Wang, W., Zhao, S., Wu, Q., Merler, S., Viboud, C., Vespignani, A., Ajelli, M., Yu, H.: Changes in contact patterns shape the dynamics of the Covid-19 outbreak in china. Science **368**(6498), 1481–1486 (2020)
44. Zhang, Z., Zeb, A., Hussain, S., Alzahrani, E.: Dynamics of Covid-19 mathematical model with stochastic perturbation. Adv. Differ. Equ. **2020**(1), 1–12 (2020)
45. Zhao, S., Lin, Q., Ran, J., Musa, S.S., Yang, G., Wang, W., Lou, Y., Gao, D., Yang, L., He, D., et al.: Preliminary estimation of the basic reproduction number of novel coronavirus (2019-ncov) in china, from 2019 to 2020: a data-driven analysis in the early phase of the outbreak. Int. J. Infect. Dis. **92**, 214–217 (2020)
46. Zhou, R., Li, F., Chen, F., Liu, H., Zheng, J., Lei, C., Wu, X.: Viral dynamics in asymptomatic patients with Covid-19. Int. J. Infect. Dis. **96**, 288–290 (2020)
47. Zhu, Y., Chen, Y.Q.: On a statistical transmission model in analysis of the early phase of Covid-19 outbreak. Stat. Biosci. 1–17 (2020)

Chapter 4
Generalized Multi-synchronization and Multi-agent Systems

The problem of *generalized multi-synchronization* (GMS) in master–slave configuration, in the context of differential algebra, can be interpreted as a leader-following consensus problem of *multi-agent system* (MAS). Here, a multi-agent system is treated as a network of interconnected systems with strictly different dynamics of the same dimension, fixed and not strongly connected. A multi-agent system is carried out to a multi-output generalized observability canonical form (MGOCF) with a family of transformations obtained from an adequate selection of the differential primitive element as a linear combination of state measurements and control inputs. This allows us to explicitly give the synchronization algebraic manifold and design a dynamic consensus protocol able to asymptotically achieve consensus for all agents in the network. Finally, an example is provided to illustrate the methodology proposed.

4.1 The Consensus Problem

Consensus problems are considered the basis of distributed computing [12]. In recent years, it has been found that is an interdisciplinary field, and the ideas behind it have been extended to numerous applications such as rendezvous, formation control, flocking, attitude alignment, sensor networks and synchronization [3, 16, 17, 19]. The main objective in these problems is that agents, with partial or complete information about their neighbors, agree upon some trajectory in common or follow a desired trajectory with a relatively constant position from each other. Some important results on consensus for nonlinear systems are given in [1, 2, 6–8, 14, 18–21]. To date, great attention is given to studying the behavior of interconnected nonlinear systems with nonidentical dynamics [5, 15, 20, 22, 26]. Some preliminary works in this direction are given in terms of the stability of sets in the practical sense.

R. Martínez-Guerra and J. P. Flores-Flores, *An Approach to Multi-agent Systems as a Generalized Multi-synchronization Problem*, Understanding Complex Systems,
https://doi.org/10.1007/978-3-031-22669-4_4

The GMS problem can be considered as a GS problem of multiple chaotic systems in master–slave configuration. Moreover, this interesting topic can be treated as well as a consensus problem. Therefore, in this book a multi-agent system is considered as a network of interconnected chaotic systems with strictly different dynamics of the same dimension. The interconnection of nodes or topology is given by a directed spanning tree with a unique leader as the root, and nodes are described by the differential equation that represents the behavior of each system. Since we treat GMS problem which appears naturally in a master–slave configuration, it is reasonable that no connections are available between followers because topology is a unidirectional type, that is, information flows only from the leader (master) to all followers (slaves). The information shared in the network is the output of the leader. From this measurement, it is possible to design a consensus protocol such that distinct trajectories of the followers that evolve in time will approach the trajectories of the leader. In this case, we say that all agent systems achieve the consensus, that is, all nodes in the network approach a common trajectory when $t \to \infty$.

The synchronization problem is attacked by the stability of the synchronization manifold that depends on the states or outputs of the systems. It is well known that for networks of interacting systems with identical dynamics there exists a trivial synchronization manifold composed of the equality between states of all systems in the network, for example, in CS. When the network has systems with strictly different dynamics the synchronization algebraic manifold is not trivial or it does not necessarily exist. Therefore, it is unclear whether this type of network can synchronize.

The GMS problem is split into two parts: first, find a mapping that relates the trajectories of the slave systems to the master system, and this generates the synchronization algebraic manifold as it will be later introduced; and second, explicitly give that mapping. Using differential-algebraic techniques, a multi-agent system is carried out to a multi-output generalized observability canonical form (MGOCF) with a family of transformations obtained from an adequate selection of the differential primitive element as a linear combination of state measurements and control inputs (the information shared in the network). This allows us to explicitly give the synchronization manifold and design a dynamic consensus protocol able to asymptotically achieve consensus for all agents in the network. This suggests that GMS problem of multiple chaotic systems can be understood as a leader-following consensus problem.

4.2 GMS, Differential Algebra and Graph Theory

Tackling the GMS problem described above requires taking into account a series of already presented concepts and some new ones. Therefore, in the following let us introduce some new definitions from differential algebra and *graph theory*.

The following definition is an extension of Definition 2.5, which will be helpful to introduce the so-called multi-output generalized observability canonical form.

Definition 4.1 *A family of systems is Picard-Vessiot (PV) if and only if the* $k_j < u >$*-vector space generated by the derivatives of the set* $\left\{\bar{y}_j^{(n_j)}, 0 \leq n_j, 1 \leq j \leq p\right\}$ *has finite dimension, with* $\bar{y}_j$ *the* j*-th differential primitive element.*

The information flow G (interaction) between $r+1$ agents considers a directed graph $G = (V, E, A)$ with a set of nodes $V = \{0, 1, \ldots, r\}$, a set of edges $E \subseteq V \times V$ and an adjacency matrix $A = [a_{ij}] \in \mathbb{R}^{(r+1)\times(r+1)}$ with nonnegative adjacency elements a_{ij} defined by

$$a_{ij} = \begin{cases} 1 & if \quad (j, i) \in E \\ 0 & \quad elsewhere \end{cases}$$

Let $L = [l_{ij}] \in \mathbb{R}^{(r+1)\times(r+1)}$ be the nonsymmetrical graph Laplacian matrix induced by the information flow G and defined as

$$l_{ij} = \begin{cases} \sum_{k=1,k\neq i}^{r+1} a_{ik} & if \quad i = j \\ \quad -a_{ij} & elsewhere \end{cases}$$

Consider a node of a graph as a dynamic agent with nonlinear dynamics given as

$$\begin{aligned} \dot{x}_{i1} &= f_{i1}^T(x_{i1}, x_{i2}, \ldots, x_{in}, u_{i1}, u_{i2}, \ldots, u_{in}), \\ \dot{x}_{i2} &= f_{i2}^T(x_{i1}, x_{i2}, \ldots, x_{in}, u_{i1}, u_{i2}, \ldots, u_{in}), \\ &\vdots \\ \dot{x}_{in} &= f_{in}^T(x_{i1}, x_{i2}, \ldots, x_{in}, u_{i1}, u_{i2}, \ldots, u_{in}) \end{aligned} \tag{4.1}$$

$\forall i \in V$, where $x_{i1} = (x_{01}, x_{11}, \ldots, x_{r1})$, $x_{i2} = (x_{02}, x_{12}, \ldots, x_{r2}), \ldots, x_{in} = (x_{0n}, x_{1n}, \ldots, x_{rn}) \in \mathbb{R}^{r+1}$, $u_{i1} = (u_{01}, \ldots, u_{r1}), \ldots, u_{in} = (u_{0n}, \ldots, u_{rn}) \in \mathbb{R}^{r+1}$.

Let x_{ij} be the value of node j of the i-th agent system $\forall i, j$ fixed. In this particular case, all interactions for nodes $x_{i1}, x_{i2}, \ldots, x_{in}$ have the same information flow G (i.e., they have the same Laplacian matrix). Thus, we can say that $x_i = (x_{i1}, \ldots, x_{in}) \in \mathbb{R}^n$ denotes the value of node $i \in V$. We define $G_x = (G, x)$ as a state of a dynamic multi-agent network (or dynamic algebraic graph) with value $x = (x_{i1}^T, \ldots, x_{in}^T)^T \in \mathbb{R}^{(r+1)n}$.

Definition 4.2 *Let* x_i *the value of node* $i \in V$. *We say nodes* i *and* j *agree if and only if* $x_i = x_j$. *In the same manner, they disagree if and only if* $x_i \neq x_j$. *Hence, the nodes of a multi-agent network have reached consensus if and only if all nodes are in agreement, or equivalently,* $x_i = x_j$, $\forall i, j \in V, i \neq j$ *[24].*

Notice that for leader-following multi-agent network systems, the consensus is reached when all nodes agree with the leader. We introduce the next definition of leader-following consensus.

Definition 4.3 *The leader-following consensus of agent systems* (4.1) *is achieved if, for each follower agent* $1 \leq l \leq r$, *there exists a differential primitive element that generates a transformation from the follower space* $\mathbb{R}^{n_l}$ *to the leader space* $\mathbb{R}^{n_0}$, *i.e.,* $H_l : \mathbb{R}^{n_l} \to \mathbb{R}^{n_0}$ *with* $H_l = \Phi_0^{-1} \circ \Phi_l$, *as well as there exists an algebraic manifold* $\mathscr{M}_l = \{(x_l, x_0) | x_0 = H_l(x_l)\}$, *a compact set* $\mathscr{B} \subset \mathbb{R}^{n_l} \times \mathbb{R}^{n_0}$, *with* $\mathscr{M}_l \subset \mathscr{B}$ *and a dynamical control law that renders this set the stable attractor of the follower agent, i.e., the trajectories beginning in* $\mathscr{B}$ *(with initial conditions* $x_i(t_0) \in \mathscr{B}, 0 \leq i \leq r$ *enter to* $\mathscr{M}_l$ *as* $t \to \infty$, *i.e.,*

$$\lim_{t \to \infty} ||H_l(x_l) - x_0|| = 0 \tag{4.2}$$

Hence, GMS problem is a leader-following consensus problem.

Definition 4.4 *The leader and all followers are in a state of GMS if the leader-following consensus of agent systems* (4.1) *is achieved.*

Definition 4.5 *Given a differentiable function* $J : \mathbb{R}^n \to \mathbb{R}$, *we denote by* ∇J *its gradient vector.*

Definition 4.6 *Let* $\mathscr{A}$ *be a closed set, and* $||x||_{\mathscr{A}}$ *denotes the distance from point* x *to a set* $\mathscr{A}$ *defined as*

$$||x||_{\mathscr{A}} := d(x, \mathscr{A}) = \inf_{\hat{x} \in \mathscr{A}} \left\{ d(\hat{x}, x) \right\}$$

where $d(x, \hat{x}) = ||x - \hat{x}||$ *is the Euclidean norm.*

Definition 4.7 *The Kronecker product of matrices* $A \in \mathbb{R}^{m \times n}$ *and* $B \in \mathbb{R}^{p \times q}$ *is defined as*

$$A \otimes B = \begin{pmatrix} a_{11} B & \dots & a_{1n} B \\ \vdots & \ddots & \vdots \\ a_{m1} B & \dots & a_{mn} B \end{pmatrix} \in \mathbb{R}^{mp \times nq}$$

Remark 4.1 *The synchronization manifold for agent systems* (4.1) *is given by*

$$\mathscr{M} = \left\{ (x_0 \otimes 1_{r \times 1}, x_l) \, | x_0 = H_1(x_1) = \cdots = H_r(x_r) \right\},$$

where $1_{r \times 1} = (1, \dots, 1)^T \in \mathbb{R}^r$, $H_l : \mathbb{R}^{n_l} \to \mathbb{R}^{n_0}$ *and* $H_l = \Phi_0^{-1} \circ \Phi_l$, $1 \leq l \leq r$. *It is said that leader-following consensus is achieved if the algebraic manifold* $\mathscr{M}$ *is a stable attractor (i.e., trajectories converge to a compact set and remain in it).*

Remark 4.2 *In a network of agents with identical dynamics, the mapping* H_l *corresponds to the identity. That is,*

$$\Phi_0 = \Phi_l$$

for $1 \leq l \leq r$*. Hence, the synchronization manifold for agent systems* (4.1) *is given as the equality of states*

$$\mathscr{M} = \{(x_0 \otimes 1_{r\times 1}, x_l) | x_0 = x_1 = \cdots = x_r\}$$

Consider the time-invariant nonlinear system

$$\dot{x} = f(x) \tag{4.3}$$

with initial condition $x(0) \in \mathbb{R}^n$, where $x(t) \in \mathbb{R}^n$ is the state vector and $f(x)$ is a Lipschitz continuous function in x. Let $x(t, x(0))$ be the solution of (4.3) at time t. We give the following definition.

Definition 4.8 *Let* $\mathscr{A}$ *be a closed invariant set,*[1] *and* $\mathscr{A}$ *is uniformly globally exponentially stable if there exist* $k > 0$ *and* $\lambda > 0$ *such that* $\forall x(0) \in \mathbb{R}^n$

$$||x(t, x(0))||_{\mathscr{A}} \leq k||x(0)||_{\mathscr{A}} e^{-\lambda t}, \quad \forall t \geq 0.$$

4.3 GMS as a Leader–Follower MAS

The synchronization problem in a master–slave configuration of multiple decoupled chaotic nonlinear systems can be reduced to a problem of leader–follower multi-agent network systems, where each agent in (4.1) is brought into a canonical form through a family of transformations given by an adequate selection of a family of differential primitive elements. Therefore, now let us introduce the following definition, which is a modification of Definition 3.1.

Definition 4.9 *A state variable* $x_{ij} \in \mathbb{R}$ *of system* (4.1) *satisfies the algebraic observability condition (AOC) if* x_{ij} *is a function of the first* $r_1, r_2 \in \mathbb{N}$ *sequential derivatives of the available output* y_i *and the input* u_i*, respectively, i.e.,*

$$x_{ij} = \phi_{ij}\left(y_i, \dot{y}_i, \ddot{y}_i, \ldots, y_i^{(r_1)}, u_i, \dot{u}_i, \ddot{u}_i, \ldots, u_i^{(r_2)}\right),$$

where $\phi_{ij} : \mathbb{R}^{(r_1+1)p} \times \mathbb{R}^{(r_2+1)m} \to \mathbb{R}$.

Consider there exist a family of elements $\bar{y}_i = (y_0, \ldots, y_r)^T \in \mathbb{R}^{r+1}$ for (4.1) with $i \in V$. Let $n \geq 0$ be the minimum integer such that $\bar{y}_i^{(n)}$ is analytically dependent on $\bar{y}_i, \bar{y}_i^{(1)}, \ldots, \bar{y}_i^{(n-1)}$:

$$\bar{H}_i\left(\bar{y}_i^{(n)}, \bar{y}_i^{(n-1)}, \ldots, \bar{y}_i, u_i^{(1)}, \ldots, u_i^{(n-1)}, u_i^{(n)}\right) = 0 \tag{4.4}$$

[1] Any solution of (4.3) that starts in $\mathscr{A}$ stays in $\mathscr{A}$ for all $t \geq 0$.

Then, the family system (4.4) can be locally solved as

$$\bar{y}_i^{(n)} = -L_i\left(\bar{y}_i^{(n-1)}, \ldots, \bar{y}_i, u_i, u_i^{(1)}, \ldots, u_i^{(n-1)}\right) + u_i^{(n)},$$

and define $z_{ik} = \bar{y}_i^{(k-1)}$, $1 \leq k \leq n$. Then we obtain a local form that can be seen as a multi-output generalized observability canonical form (MGOCF):

$$\begin{aligned}
\dot{z}_{i1} &= z_{i2}, \\
\dot{z}_{i2} &= z_{i3}, \\
&\vdots \\
\dot{z}_{in-1} &= z_{in}, \\
\dot{z}_{in} &= -\bar{L}\left(z_{i1}, \ldots, z_{in}, u_i, u_i^{(1)}, \ldots, u_i^{(n-1)}\right) + u_i^{(n)}, \\
\bar{y}_i &= z_{i1}
\end{aligned} \tag{4.5}$$

with $\bar{L}_i(\cdot) = \left(\bar{L}_0(\cdot), \bar{L}_1(\cdot), \ldots, \bar{L}_r(\cdot)\right)^T \in \mathbb{R}^{r+1}$, $u_i = (u_0, u_1, \ldots, u_r)^T \in \mathbb{R}^{r+1}$, $z_{i1} = (z_{01}, \ldots, z_{r1})^T, \ldots, z_{in} = (z_{0n}, \ldots, z_{rn})^T \in \mathbb{R}^{r+1}$. Now, we can establish the next lemma:

Lemma 4.1 *The dynamic agent systems* (4.1) *are transformable to a MGOCF if and only if they are a PV family.*

Proof (Sufficiency) Let us consider the set $\left\{\xi_i, \xi_i^{(1)}, \ldots, \xi_i^{(n-1)}\right\}$ be a finite differential transcendence basis with $\xi_i^{(k-1)} = y_i^{(k-1)}$, $0 \leq i \leq r$, $1 \leq k \leq n$ with $n \geq 0$ the minimum integer such that $y_i^{(n)}$ is dependent on $y_i, y_i^{(1)}, \ldots, y_i^{(n-1)}, u_i, \ldots$. Redefining $\xi_{ik} = \xi_i^{(k-1)}$, $i \leq k \leq n$, it is obtained

$$\begin{aligned}
\dot{\xi}_{ij} &= \xi_{ij+1}, \quad 1 \leq j \leq n-1, \\
\dot{\xi}_{in} &= -\bar{L}_i\left(\xi_{i1}, \ldots, \xi_{in}, u_i, u_i^{(1)}, \ldots, u_i^{(n-1)}\right) + u_i^{(n)}, \\
y_i &= \xi_{i1}
\end{aligned}$$

(Necessity) It is immediate. □

Remark 4.3 *We can choose the differential primitive element as*

$$y_i = \sum_{\bar{j}} \alpha_{i\bar{j}} x_{i\bar{j}} + \sum_{\bar{l}} \beta_{i\bar{l}} u_{i\bar{l}}, \quad \alpha_{i\bar{j}}, \beta_{i\bar{l}} \in k < u_i >, \tag{4.6}$$

where K $< u_i >$ is a differential field generated by K and u_i and their differential quantities.

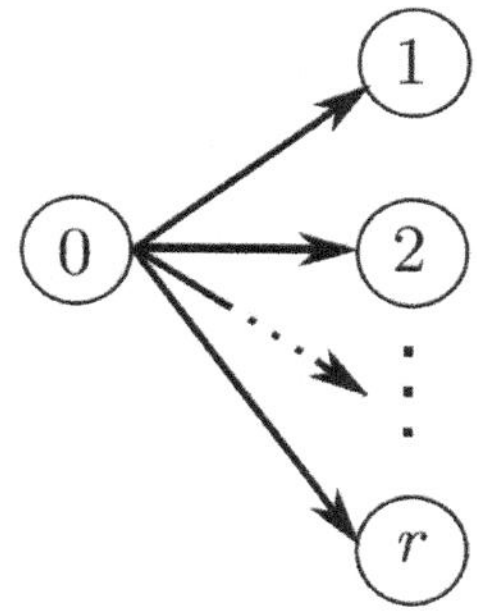

Fig. 4.1 MAS in a leader-following configuration (unique leader as root), directed spanning tree $G = (V, E, A)$

Let $G = (V, E, A)$ be a directed graph (the information flow) between $r + 1$ agents systems in form (4.1) in a leader-following configuration such that node 0 denotes the leader and l for $1 \leq l \leq r$ are the corresponding follower nodes as given in Fig. 4.1.

Remark 4.4 *We assigned one fixed interaction topology associated with the $r + 1$ agents, i.e., each state of the state vector x_i in* (4.1) *is associated with the same dynamic directed graph G, as well as interaction topology is not strongly connected (i.e., there is no sequence of edges in the directed graph from every node to every other nodes) and is time-invariant. Thus, the information flow between the $r + 1$ agents is fixed for all time. From Fig. 4.1, the adjacency and Laplacian matrix associated with G are given by*

$$A = \begin{bmatrix} 0 & 0 & \dots & 0 \\ 1 & 0 & \dots & 0 \\ \vdots & \vdots & \ddots & \vdots \\ 1 & 0 & \dots & 0 \end{bmatrix} \in \mathbb{R}^{(r+1)\times(r+1)} \tag{4.7}$$

and

$$L = \begin{bmatrix} 0 & 0 & \dots & 0 \\ -1 & 1 & \dots & 0 \\ \vdots & \vdots & \ddots & \vdots \\ -1 & 0 & \dots & 0 \end{bmatrix} \in \mathbb{R}^{(r+1)\times(r+1)}, \tag{4.8}$$

respectively.

Above process is shown in Fig. 4.2. We can take MAS (4.1) to its MGOCF through transformations Φ_i obtained from the differential primitive element and its time derivatives. In what follows we choose an adequate dynamical control law that imposes similar dynamics to the follower systems.

Theorem 4.1 *Let us consider $r + 1$ agent systems of the form* (4.1) *in a leader-following configuration with interaction topology given by the directed graph G such*

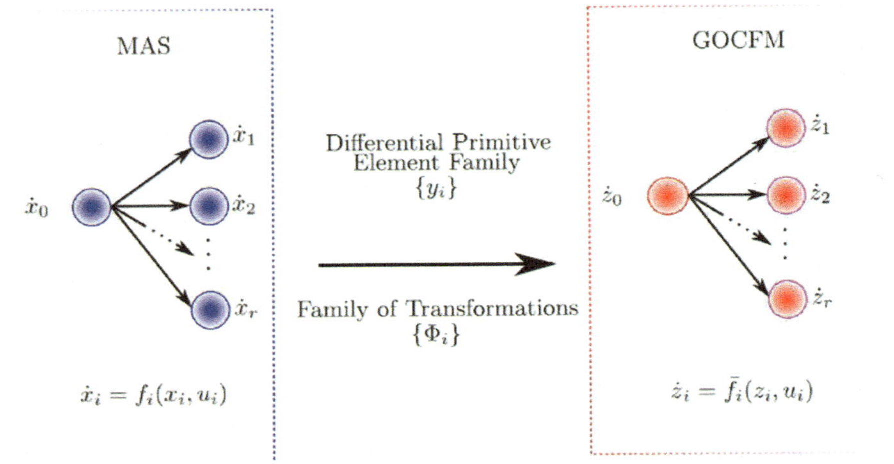

Fig. 4.2 Multi-agent systems in generalized multi-synchronization

that MAS can be transformable to a MGOCF and by defining $z_i = (z_{i1}, z_{i2}, \ldots, z_{in})$ *as the trajectories of the leader and the r followers in transformed coordinates, respectively, with* $z_{ij} = y_i^{(j-1)}$ *for* $0 \leq i \leq r$ *and* $1 \leq j \leq n$*. Choosing a dynamic protocol for the l-th follower as follows:*

$$u_l^\gamma = -\bar{L}_0 = (z_{01}, \ldots, z_{0n}) + \bar{L}_1 \left(z_{l1}, \ldots, z_{ln}, u_l, \ldots, u_l^{(n-1)}\right) - a_{l1} \sum_{q=1}^{n} k_{lq} \left(z_{0q}, -z_{1q}\right) \tag{4.9}$$

where k_{lj} *are positive gains and* a_{lj} *are elements of the adjacency matrix* (4.7)*. Then the leader-following consensus problem is solved if*

$$\lim_{t\to\infty} ||z_0 - z_l|| = 0, \tag{4.10}$$

The last leads to

$$\lim_{t\to\infty} ||H_l(x_l) - x_0|| = 0, \tag{4.11}$$

for all $1 \leq l \leq r$*, with* y_i *a differential primitive elements family for the* $r+1$ *dynamic agents.*

Proof We can choose $u_0 = 0 \in \mathbb{R}$, without loss of generality, for the leader. Consider differential primitive elements for $i \in V$ as the available outputs for agent systems (4.1) given by

$$y_i = \sum_j \alpha_{ij} x_{ij} + \sum_k \beta_{ik} u_{ik} = z_{i1}, \quad \alpha_{ij}, \beta_{ik} \in \mathbb{R} < u_i > . \tag{4.12}$$

From Lemma 4.1 and (4.12) we obtain

$$\begin{aligned} \dot{z}_{ij} &= z_{ij+1}, \quad 0 \le i \le r, 1 \le j \le n-1, \\ \dot{z}_{in} &= -\bar{L}_i \left(z_{il}, \ldots, z_{in}, u_i, u_i^{(1)}, \ldots, u_i^{(n-1)} \right) + u_i^{(n)}. \end{aligned} \tag{4.13}$$

We construct control signals as a chain of integrators, i.e., $u_{i1} = u_i, u_{i2} = \dot{u}_i, \ldots, u_{i\gamma} = u_i^{(\gamma-1)}$, using dynamical controller (4.9), then system (4.13) can be written as the closed-loop system:

$$\dot{z} = \begin{pmatrix} 0_{(r+1)} & I_{(r+1)} & \ldots & 0_{(r+1)} \\ \vdots & \vdots & \ddots & \vdots \\ 0_{(r+1)} & 0_{(r+1)} & \ldots & I_{(r+1)} \\ -K_1 L & -K_2 L & \ldots & -K_n L \end{pmatrix} z + \begin{pmatrix} 0_{(r+1)\times 1} \\ \vdots \\ 0_{(r+1)\times 1} \\ \bar{L}_0 \cdot 1_{(r+1)\times 1} \end{pmatrix}$$

where $z = (z_{01}, \ldots, z_{0n}, \ldots, z_{r1}, \ldots, z_{rn})^T \in \mathbb{R}^{(r+1)n}$. Laplacian matrix L is defined as in (4.8) and

$$K_j = \begin{pmatrix} 0 & 0 & \ldots & 0 & 0 \\ 0 & k_{1j} & \ldots & 0 & 0 \\ \vdots & \vdots & \ddots & \vdots & \vdots \\ 0 & 0 & \ldots & k_{(r-1)j} & 0 \\ 0 & 0 & \ldots & 0 & k_{rj} \end{pmatrix} \in \mathbb{R}^{(r+1)\times(r+1)} \tag{4.14}$$

for $1 \le j \le n$. Let $e_z = z_0 \otimes 1_{r\times 1} - z_l$ be the synchronization error, then synchronization error dynamics is the following augmented system

$$\begin{aligned} \dot{e}_{zlj} &= e_{zlj+1}, \quad i \le j \le n-1, 1 \le l \le r, \\ e_{zln} &= -\bar{L}_0 (z_{01}, \ldots, z_{0n}) \cdot 1_{r\times 1} + \bar{L}_l (z_{l1}, \ldots, z_{ln}, u_{l1}, \ldots, u_{ln}) - \dot{u}_{ln}, \\ \dot{u}_{ln} &= u_{lk+1}, \quad 1 \le k \le n-1, \\ \dot{u}_{l\gamma} &= -\bar{L}_0 (z_{01}, \ldots, z_{0n}) \cdot 1_{r\times 1} + \bar{L}_l (z_{l1}, \ldots, z_{ln}, u_{l1}, \ldots, u_{ln}) + \bar{M} e_z, \end{aligned} \tag{4.15}$$

with $\bar{L}_l(\cdot) = \left(\bar{L}_1(\cdot), \ldots, \bar{L}_r(\cdot)\right)^T \in \mathbb{R}^r$, $e_z = (e_{z11}, \ldots, e_{z1n}, \ldots, e_{zr1}, \ldots, e_{zrn})^T$ and $\bar{M} = (M_1, \ldots, M_n) \in \mathbb{R}^{r\times rn}$.

From (4.15) we finally have that

$$\dot{e}_z = \Xi e_z, \tag{4.16}$$

with

$$\Xi = \begin{pmatrix} 0_{r\times r} & I_{r\times r} & 0_{r\times r} & \dots & 0_{r\times r} & 0_{r\times r} \\ 0_{r\times r} & 0_{r\times r} & I_{r\times r} & \dots & 0_{r\times r} & 0_{r\times r} \\ \vdots & \vdots & \vdots & \ddots & \vdots & \vdots \\ 0_{r\times r} & 0_{r\times r} & 0_{r\times r} & \dots & I_{r\times r} & 0_{r\times r} \\ 0_{r\times r} & 0_{r\times r} & 0_{r\times r} & \dots & 0_{r\times r} & I_{r\times r} \\ -M_1 & -M_2 & -M_3 & \dots & -M_{n-1} & -M_n \end{pmatrix} \in \mathbb{R}^{rn\times rn} \tag{4.17}$$

and the control gain matrices M_j given by

$$M_j = \begin{pmatrix} k_{1j} & 0 & \dots & 0 & 0 \\ 0 & k_{2j} & \dots & 0 & 0 \\ \vdots & \vdots & \ddots & \vdots & \vdots \\ 0 & 0 & \dots & k_{(r-1)j} & 0 \\ 0 & 0 & \dots & 0 & k_{rj} \end{pmatrix} \in \mathbb{R}^{r\times r}, \tag{4.18}$$

with positive constants k_{lj} which are chosen such that matrix $\Xi \in \mathbb{R}^{rn\times rn}$ is a Hurwitz matrix. □

Theorem 4.2 *Consider the dynamical protocol* (4.9). *Let* T^* *be the synchronization time. Then the synchronization algebraic manifold for agent systems* (4.1) *given by*

$$\mathscr{M} = \{(x_0 \otimes 1_{r\times r}, x_l) | x_0 = H_1(x_1) = \cdots = H_r(x_r)\}, \forall t \geq T^*$$

is uniformly globally exponentially stable, where $H_l : \mathbb{R}^{n_l} \to \mathbb{R}^{n_0}$ *with* $H_l = \Phi_0^{-1} \circ \Phi_l$, $\mathbb{R}^{n_l}$ *and* $\mathbb{R}^{n_0}$ *denote the follower and leader space, respectively,* $1 \leq l \leq r$.

Proof Assume $z_0 \otimes 1_r$, $z_l = (z_{l1}, \dots, z_{lr})^T \in \mathbb{R}^{rn\times 1}$, $1_{r\times 1} := (1 \dots 1)^T \in \mathbb{R}^{r\times 1}$. Define the synchronization algebraic manifold in transformed coordinates as follows

$$\mathscr{M}_z := \{(z_0 \otimes 1_{r\times 1}, z_l) \,|z_0 \otimes 1_{r\times 1} = z_l\}.$$

The distance function $||\cdot||_{\mathscr{M}_z}$ is

$$\begin{aligned} ||\,(z_0 \otimes 1_{r\times 1}, z_l)\,||_{\mathscr{M}_z} &= \inf_{(\xi_0\otimes 1_{r\times 1}, \xi_l)\in \mathscr{M}_z} \| \begin{pmatrix} z_0 \otimes 1_{r\times 1} - \xi_0 \otimes 1_{r\times 1} \\ z_l - \xi_l \end{pmatrix} \|, \\ &= \inf_{\xi_0 \otimes 1_{r\times 1}\in \mathbb{R}^{rn}} \| \begin{pmatrix} z_0 \otimes 1_{r\times 1} - \xi_0 \otimes 1_{r\times 1} \\ z_l - \xi_0 \otimes 1_{r\times 1} \end{pmatrix} \|, \\ &= \inf_{\xi_0 \otimes 1_{r\times 1}\in \mathbb{R}^{rn}} J(\xi_0), \end{aligned}$$

where

$$J(\xi_0) = \sqrt{\| z_0 \otimes 1_{r\times 1} - \xi_0 \otimes 1_{r\times 1} \|^2 + \| z_l - \xi_0 \otimes 1_{r\times 1} \|^2}.$$

Due to $J(\xi_0)$ is a continuous function, $J(\xi_0)$ attains its infimum,

$$\| (z_0 \otimes 1_{r\times 1}, z_l) \|_{\mathscr{M}_z} = \min_{\xi_0 \otimes 1_{r\times 1} \in \mathbb{R}^{rn}} J(\xi_0).$$

We obtain the $argmin J(\xi_0)$ from $\nabla J(\xi_0) = 0$,

$$\nabla J(\xi_0) = \frac{1}{J(\xi_0)} \left((z_0 \otimes 1_{r\times 1} - \xi_0 \otimes 1_{r\times 1})^T + (z_l - \xi_0 \otimes 1_{r\times 1})^T \right)$$

Making $\nabla J(\xi_0) = 0$ this yields to

$$argmin J(\xi_0) = \xi_0 \otimes 1_{r\times 1} = \frac{1}{2} (z_0 \otimes 1_{r\times 1} + z_l).$$

Finally, $\| \cdot \|_{\mathscr{M}_z}$ is given by

$$\begin{aligned}
\| (z_0 \otimes 1_{r\times 1}, z_l) \|_{\mathscr{M}_z} &= \left\| \begin{pmatrix} z_0 \otimes 1_{r\times 1} - \xi_0 \otimes 1_{r\times 1} \\ z_l - \xi_0 \otimes 1_{r\times 1} \end{pmatrix} \right\|, \\
&= \frac{1}{2} \left\| \begin{pmatrix} z_0 \otimes 1_{r\times 1} - z_l \\ -(z_0 \otimes 1_{r\times 1} - z_l) \end{pmatrix} \right\|, \\
&= \frac{1}{\sqrt{2}} \| z_0 \otimes 1_{r\times 1} - z_l \|,
\end{aligned}$$

that is equivalent to

$$\| z_0 \otimes 1_{r\times 1} - z_l \| = \sqrt{2} \, \| (z_0 \otimes 1_{r\times 1}, z_l) \|_{\mathscr{M}_z} \tag{4.19}$$

Since Ξ is a Hurwitz stable matrix, (4.16) has a unique solution over $[0, t_1]$ where t_1 can be arbitrarily large; this implies that system (4.16) has a finite time escape detectable through $\| (z_0 \otimes 1_{r\times 1}, z_l) \|_{\mathscr{M}_z}$ (solution e_z does not escape to infinity at a finite time). Moreover, the synchronization error $e_z = z_0 \otimes 1_{r\times 1} - z_l$ is asymptotically stable, then the Lyapunov equation

$$P\Xi + \Xi^T P = -Q$$

is satisfied for $P > 0, Q > 0$. Let $V = e_z^T P e_z$ be the Lyapunov candidate function, V satisfies the following inequalities:

$$2\lambda_{min}(P) \| (z_0 \otimes 1_{r\times 1}, z_l) \|^2_{\mathscr{M}_z} \leq V \leq 2\lambda_{max}(P) \| (z_0 \otimes 1_{r\times 1}, z_l) \|^2_{\mathscr{M}_z}. \tag{4.20}$$

Differentiating V with respect to the trajectories of (4.16) we obtain

$$\dot{V} = e_z^T(P\Xi + \Xi^T P)e_z = -e_z^T Q e_z,$$
$$= -\lambda_{min}(Q) \parallel e_z \parallel^2,$$

thus

$$\dot{V} \leq -2\lambda_{min}(Q) \parallel (z_0 \otimes 1_{r\times 1}, z_l) \parallel^2_{\mathscr{M}_z} . \tag{4.21}$$

From (4.20) and (4.21) using Theorem A.10 given in [9]. Then $\mathscr{M}_z$ is uniformly globally exponentially stable. It is easy to verify from (4.20) and (4.21) the following relationship

$$\dot{V}\left(z_0(t) \otimes 1_{r\times 1}, z_l(t)\right) \leq -\beta V\left(z_0(t) \otimes 1_{r\times 1}, z_l(t)\right), \tag{4.22}$$

where $\beta := \frac{\lambda_{min}(Q)}{\lambda_{max}(P)}$; from comparison lemma, the solution of differential inequality (4.22) is bounded (see [4]),

$$V(z_0(t) \otimes 1_{r\times 1}, z_l(t)) \leq e^{-\beta t} V(z_0(t) \otimes 1_{r\times 1}, z_l(t)), \tag{4.23}$$

From (4.20) and (4.23) the next exponential estimate is fulfilled:

$$\parallel (z_0(t) \otimes 1_{r\times 1}, z_l(t)) \parallel_{\mathscr{M}_z} \leq \alpha e^{-\frac{\beta}{2}t} \parallel (z_0(0) \otimes 1_{r\times 1}, z_l(0)) \parallel_{\mathscr{M}_z} \tag{4.24}$$

with $\alpha := \sqrt{\frac{\lambda_{max}(P)}{\lambda_{min}(P)}}$.

On the other hand note that for $t \geq T^*$, $z_0(T^*) \otimes 1_{r\times 1} = z_l(T^*)$, hence for each l fixed, $1 \leq l \leq r$,

$$\Phi_0(x_0(T^*)) = z_0(T^*) = z_l(T^*) = \Phi_l(x_l(T^*))$$

then

$$\begin{aligned} \Phi_0^{-1} \circ \Phi_l(x_l(T^*)) &= \Phi_0^{-1}(z_l(T^*)), \\ &= \Phi_0^{-1}(z_0(T^*)), \\ &= \Phi_0^{-1} \circ \Phi_0(x_0(T^*)), \\ &= x_0(T^*). \end{aligned}$$

Therefore,

$$x_l(T^*) = \Phi_0^{-1}(z_l(T^*)) = \Phi_0^{-1}(z_0(T^*)) = x_0(T^*),$$

that immediately implies

$$\mathscr{M} = \{(x_0 \otimes 1_{r\times 1}, x_l)\,|x_0 \otimes 1_{r\times 1} = x_l\} = \mathscr{M}_z, \quad \forall t \geq T^*.$$

Finally, we obtain from (4.24) the exponential estimate for $\mathscr{M}$ for all $t \geq T^*$. This ends the proof. □

4.4 Numerical Example of GMS

Let us consider three different chaotic systems in a leader-following configuration with directed graph $G = (V, E, A)$ shown in Fig. 4.2. The leader node 0 or Colpitts system is given by:

$$\Sigma_{colpitts} = \begin{cases} \dot{x}_{01} = & -a_0 \exp(-x_{02}) + a_0 x_{03} + a_0, \\ \dot{x}_{02} = & b_0 x_{03}, \\ \dot{x}_{03} = & -c_0 x_{01} - c_0 x_{02} - d_0 x_{03}. \end{cases} \tag{4.25}$$

where $x_0 = (x_{01}, x_{02}, x_{03})^T \in \mathbb{R}^{n_0}$, $n_0 = 3$, with $a_0 = b_0 = 6.2723$, $c_0 = 0.0797$, $d_0 = 0.6898$ and initial conditions $x_0(0) = (0.6, 0.1, -0.6)^T$ such that (4.25) has a chaotic behavior (see Fig. 4.3) and consider its available output to be $y_0 = x_{02}$.

System (4.25) satisfies the AOC, that is to say, the following relations are satisfied:

$$\Phi_0^{-1} := \begin{pmatrix} x_{01} \\ x_{02} \\ x_{03} \end{pmatrix} = \begin{pmatrix} \frac{1}{b_0 c_0}(-z_{03} - d_0 z_{02} - b_0 c_0 z_{01}) \\ z_{01} \\ \frac{1}{b_0} z_{02} \end{pmatrix} \tag{4.26}$$

with $z_{01} = y_0$, $z_{02} = \dot{y}_0$ and $z_{03} = \ddot{y}_0$.

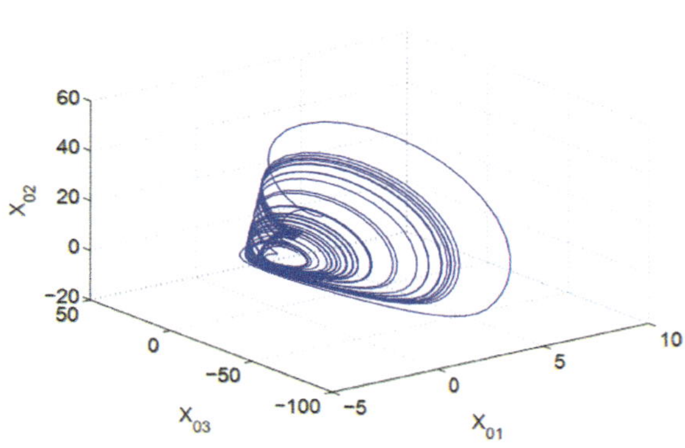

Fig. 4.3 Phase portrait of Colpitts system

Fig. 4.4 Phase portrait of Rössler system

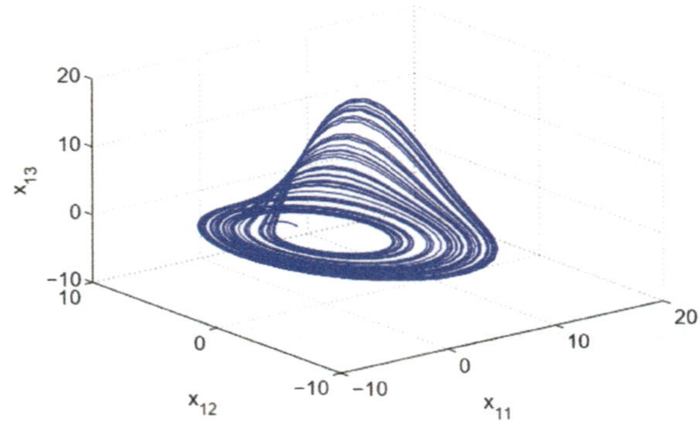

Let the first follower or node 1 be the Rössler system given by:

$$\Sigma_{Rossler} = \begin{cases} \dot{x}_{11} = & -(x_{12} + x_{13}), \\ \dot{x}_{12} = & x_{11} + a_1 x_{12}, \\ \dot{x}_{13} = & b_1 + x_{13}(x_{11} - c_1). \end{cases} \tag{4.27}$$

where $x_1 = (x_{11}, x_{12}, x_{13})^T \in \mathbb{R}^{n_1}$, $n_1 = 3$, with constant parameters $a_1 = b_1 = 0.2$, $c_1 = 5$ and initial conditions $x_0(0) = (1, 2, -5)^T$ such that (4.27) exhibits chaos (Fig. 4.4). We choose $y_1 = x_{12} + u_{11}$ be its available output such that the AOC is satisfied. The function Φ_1 that transforms original coordinates of system (4.27) into transformed coordinates is given by:

$$\Phi_1 := \begin{pmatrix} z_{11} \\ z_{12} \\ z_{13} \end{pmatrix} = \begin{pmatrix} x_{12} + u_{11} \\ x_{11} + a_1 x_{12} + u_{12} \\ a_1 x_{11} + (a_1^2 - 1)x_{12} - x_{13} + u_{13} \end{pmatrix} \tag{4.28}$$

with $z_{11} = y_1$, $z_{12} = \dot{z}_{11}$ and $z_{13} = \dot{z}_{12}$.

And let us consider the second follower or node 2 as a Chua chaotic system:

$$\Sigma_{Chua} = \begin{cases} \dot{x}_{21} = a_2(x_{22} - x_{21} - \nu_{x_2}), \\ \dot{x}_{22} = \quad x_{21} - x_{22} + x_{23}, \\ \dot{x}_{23} = \quad -b_2 x_{22}. \end{cases} \tag{4.29}$$

with

$$\nu_{x_2} = m_0 x_{21} + \frac{1}{2}(m_1 - m_0)(|x_{21} + 1| - |x_{21} - 1|), \tag{4.30}$$

where $x_2 = (x_{21}, x_{22}, x_{23})^T \in \mathbb{R}^{n_2}$, $n_2 = 3$, with constant parameters $a_2 = 15$, $b_2 = 25.58$, $m_0 = -5/7$, $m_1 = 8/7$ and initial conditions $x_0(0) = (0.6, 0.1, -0.6)^T$ such that (4.29) is chaotic and is shown in Fig. 4.5. Considering $y_2 = x_{23} + u_{21}$ as its avail-

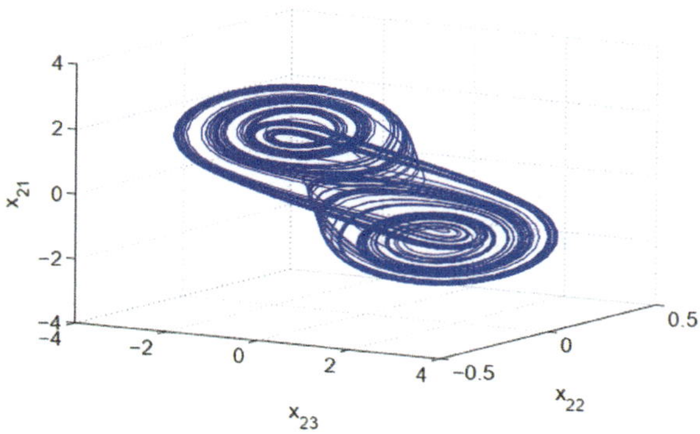

Fig. 4.5 Phase portrait of Chua system

able output such that the AOC condition is fulfilled. The function Φ_2 that transforms original coordinates of system (4.27) into transformed coordinates is given by:

$$\Phi_2 := \begin{pmatrix} z_{21} \\ z_{22} \\ z_{23} \end{pmatrix} = \begin{pmatrix} x_{23} + u_{21} \\ -b_2 x_{22} + u_{22} \\ -b_2(x_{21} - x_{22} + x_{23}) + u_{23} \end{pmatrix}, \tag{4.31}$$

with $z_{11} = y_2$, $z_{22} = \dot{z}_{21}$ and $z_{23} = \dot{z}_{22}$.

Consider the following augmented system (Theorem 4.1):

$$\begin{aligned} \dot{x}_{01} &= -a_0 \exp(-x_{02}) + a_0 x_{03} + a_0, \\ \dot{x}_{11} &= -(x_{12} + x_{13}), \\ \dot{x}_{21} &= a_2(x_{22} - x_{21} - \nu_{x_2}), \\ \dot{x}_{02} &= b_0 x_{03}, \\ \dot{x}_{12} &= x_{11} + a_1 x_{12}, \\ \dot{x}_{22} &= x_{21} - x_{22} + x_{23}, \\ \dot{x}_{03} &= -c_0 x_{01} - c_0 x_{02} - d_0 x_{03}, \\ \dot{x}_{13} &= b_1 + x_{13}(x_{11} - c_1), \\ \dot{x}_{23} &= -b_2 x_{22}, \end{aligned} \tag{4.32}$$

and choose the differential primitive elements as

$$\begin{aligned} y_0 &= x_{02}, \\ y_1 &= x_{12} + u_{11}, \\ y_2 &= x_{23} + u_{21}. \end{aligned} \tag{4.33}$$

From (4.32) and (4.33) we obtain a MGOCF

$$
\begin{aligned}
\dot{z}_{01} &= z_{02},\\
\dot{z}_{11} &= z_{12},\\
\dot{z}_{21} &= z_{22},\\
\dot{z}_{02} &= z_{03},\\
\dot{z}_{12} &= z_{13},\\
\dot{z}_{22} &= z_{23},\\
\dot{z}_{03} &= a_0 b_0 c_0 \exp(-z_{01}) + (-a_0 c_0 - b_0 c_0 + d_0^2 - d_0) z_{02}\\
&\quad - d_0 z_{03} - a_0 b_0 c_0 = \varphi_0(z_0 j),\\
\dot{z}_{13} &= -c_1(z_{11} - u_{11}) + (a_1 c_1 - 1)(z_{12} - u_{12})\\
&\quad + (a_1 - c_1)(z_{13} - u_{13}) - (-(z_{11} - u_{11}) + a_1(z_{12} - u_{12}) - (z_{13} - u_{13}))\\
&\quad \times (z_{12} - u_{12} - a(z_{11} - u_{11})) - b + \dot{u}_{13} = \varphi(z_{1j}, u_{1j}) + \dot{u}_{13},\\
\dot{z}_{23} &= a_2 b_2(u_{21} - z_{21}) + b_2(u_{22} - z_{22}) + (a_2 + 1)(u_{23} - z_{23})\\
&\quad + a_2 b_2 v_{z_2} + \dot{u}_{23} = \varphi(z_{2j}, u_{2j}) + \dot{u}_{23},
\end{aligned}
\tag{4.34}
$$

with

$$
\begin{aligned}
v_{z_2} &= m_0\left(u_{21} - z_{21} + \frac{1}{b}(u_{22} - z_{22} + u_{23} - z_{23})\right) + \frac{1}{2}(m_1 - m_0)\\
&\quad \times \left(\| u_{21} - z_{21} + \frac{1}{b}(u_{22} - z_{22} + u_{23} - z_{23}) + 1 \| \right.\\
&\quad \left. - \| u_{21} - z_{21} + \frac{1}{b}(u_{22} - z_{22} + u_{23} - z_{23}) - 1 \| \right),
\end{aligned}
$$

then the dynamical controller is given by

$$
\begin{aligned}
\dot{u}_{11} &= u_{12},\\
\dot{u}_{21} &= u_{22},\\
\dot{u}_{12} &= u_{13},\\
\dot{u}_{22} &= u_{23},\\
\dot{u}_{13} &= \bar{u}_1,\\
\dot{u}_{23} &= \bar{u}_2.
\end{aligned}
$$

Since, the objective is $|z_{01} - z_{11}| \to 0$, $|z_{01} - z_{21}| \to 0, \ldots$, $|z_{03} - z_{13}| \to 0$, $|z_{03} - z_{23}| \to 0$ as $t \to \infty$. Then, the synchronization error vector is

$$\begin{aligned}
e_{z11} &= z_{01} - z_{11},\\
e_{z21} &= z_{01} - z_{21},\\
e_{z12} &= z_{02} - z_{12},\\
e_{z22} &= z_{02} - z_{22},\\
e_{z13} &= z_{03} - z_{13},\\
e_{z23} &= z_{03} - z_{23},
\end{aligned}$$

and this yields to

$$\begin{aligned}
\dot{e}_{z11} &= e_{z12},\\
\dot{e}_{z21} &= e_{z22},\\
\dot{e}_{z12} &= e_{z13},\\
\dot{e}_{z22} &= e_{z23},\\
\dot{e}_{z13} &= \varphi(z_{0j}) - \varphi_1(z_{1j}, u_{1j}) - \bar{u}_1,\\
\dot{e}_{z23} &= \varphi(z_{0j}) - \varphi_2(z_{2j}, u_{2j}) - \bar{u}_2,
\end{aligned} \tag{4.35}$$

and now choose

$$\begin{aligned}
\bar{u}_1 &= \varphi_0 - \varphi_1 + \sum_{q=1}^{n} a_{21} k_{1q}(z_{0q} - z_{1q}),\\
\bar{u}_2 &= \varphi_0 - \varphi_2 + \sum_{q=1}^{n} a_{31} k_{2q}(z_{0q} - z_{2q}),
\end{aligned}$$

where a_{ij} are elements of the adjacency matrix A in (4.7) and k_{ij} are positive gain constants.

Hence, error dynamics (4.35) could be rewritten as

$$\dot{e}_z = \Xi e_z \tag{4.36}$$

with

$$\Xi = \begin{pmatrix}
0 & 0 & 1 & 0 & 0 & 0\\
0 & 0 & 0 & 1 & 0 & 0\\
0 & 0 & 0 & 0 & 1 & 0\\
0 & 0 & 0 & 0 & 0 & 1\\
-k_{11} & 0 & -k_{12} & 0 & -k_{13} & 0\\
0 & -k_{21} & 0 & -k_{22} & 0 & -k_{23}
\end{pmatrix}$$

and defining

$$\mathrm{K}_j = \begin{pmatrix} k_{1j} & 0 \\ 0 & k_{2j} \end{pmatrix} \in \mathbb{R}^{2\times 2},$$

as gain constant matrices for $j = 1, 2, 3$.

Due to information flow G, (4.36) and (4.34), the closed-loop system takes the following form

$$\dot{z} = \begin{pmatrix} 0_{3\times 3} & I_{3\times 3} & 0_{3\times 3} \\ 0_{3\times 3} & 0_{3\times 3} & I_{3\times 3} \\ -M_1 L & -M_2 L & -M_3 L \end{pmatrix} z + \begin{pmatrix} 0_{3\times 1} \\ 0_{3\times 1} \\ \varphi_0 \cdot 1_{3\times 1} \end{pmatrix}, \tag{4.37}$$

where L is the Laplacian in (4.8) and gain matrices

$$M_j = \begin{pmatrix} 0 & 0_{1\times 2} \\ 0_{2\times 1} & \mathrm{K}_j \end{pmatrix} \in \mathbb{R}^{3\times 3}$$

The consensus is achieved for all agents in network (4.37) if gain matrices K_i are chosen such that matrix Ξ in (4.36) is Hurwitz stable. Let us consider the matrix gains as $\mathrm{K}_1 = 500 I_{2\times 2}, \mathrm{K}_2 = 600 I_{2\times 2}$ and $\mathrm{K}_3 = 700 I_{2\times 2}$. Hence Fig. 4.6 shows that leader-following consensus is achieved in transformed coordinates, i.e., the relationship (4.10) is satisfied.

Figure 4.7 describes that leader-following consensus is achieved in original coordinates through the inverse transformations (4.11), and $\Phi_0^{-1} \circ \Phi_l$ and $\Phi_0^{-1} \circ \Phi_2$ are obtained from vector functions (4.26) and (4.31), that is,

$$\begin{aligned} H_1(x_1(t)) &:= \Phi_0^{-1} \circ \Phi_1(x_1(t)), \\ &= \begin{bmatrix} -\frac{1}{b_0 c_0}\left(-a_1 x_1 1 - (a^2 - 1)x_{12} + x_{13} - u_{13} - d_0(x_{11} + a_1 x_{12} + u_{12})\right) - (x_{12} + u_{11}) \\ x_{12} + u_{11} \\ \frac{x_{11} + a_1 x_{12} u_{12}}{b_0} \end{bmatrix} \end{aligned}$$

and

$$\begin{aligned} H_2(x_2(t)) &:= \Phi_0^{-1} \circ \Phi_2(x_1(t)), \\ &= \begin{bmatrix} -\frac{1}{b_0 c_0}(-b_2(x_{21} - x_{22} + x_{23}) + u_{23} - d_0(-b_2 x_{22} + u_{22})) - (x_{23} - u_{21}) \\ x_{23} + u_{21} \\ \frac{-b x_{22} + u_{22}}{b_0} \end{bmatrix} \end{aligned}$$

In this case, the synchronization algebraic manifold is given by:

$$\mathscr{M} = \{(x_0 \otimes 1_{r\times 1}, x_l)\,|x_0 = H(x_1) = H(x_2)\}$$

with $x_l = (x_{11}, x_{21}, x_{12}, x_{22}, x_{13}, x_{23})^T$.

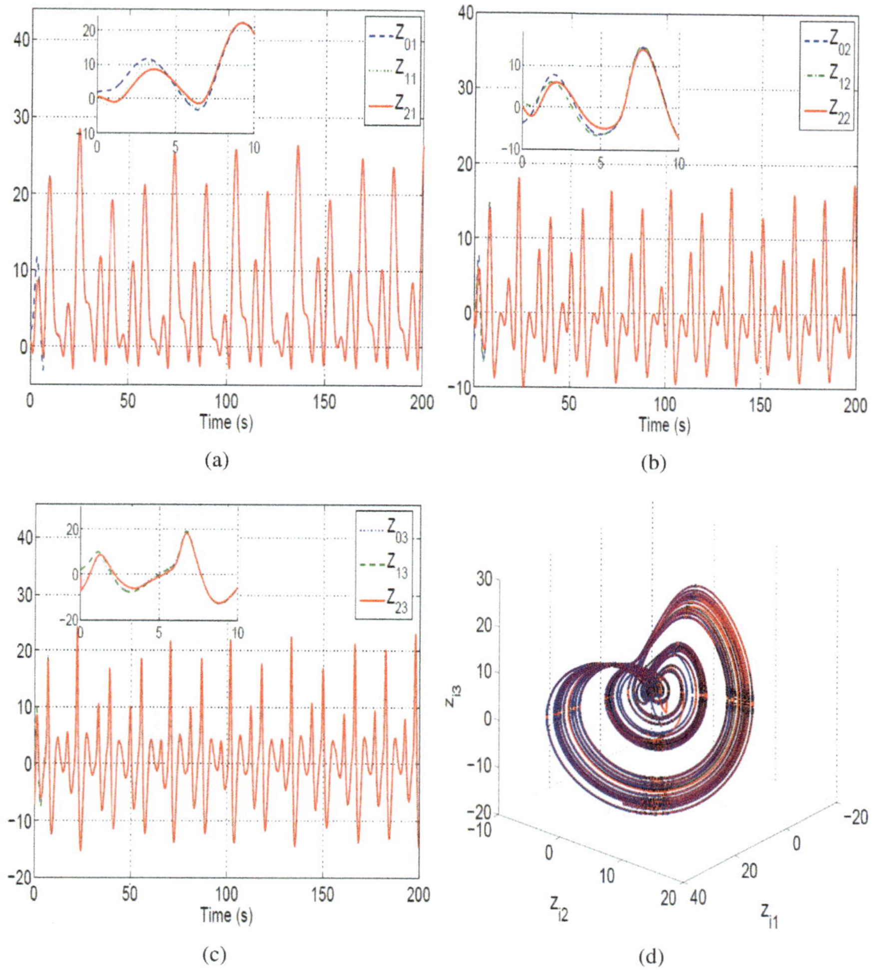

Fig. 4.6 Agreement in transformed coordinates. **a–c** Show each state trajectory in the transformed coordinates where, clearly, all systems follow the same path. **d** Shows the phase portrait

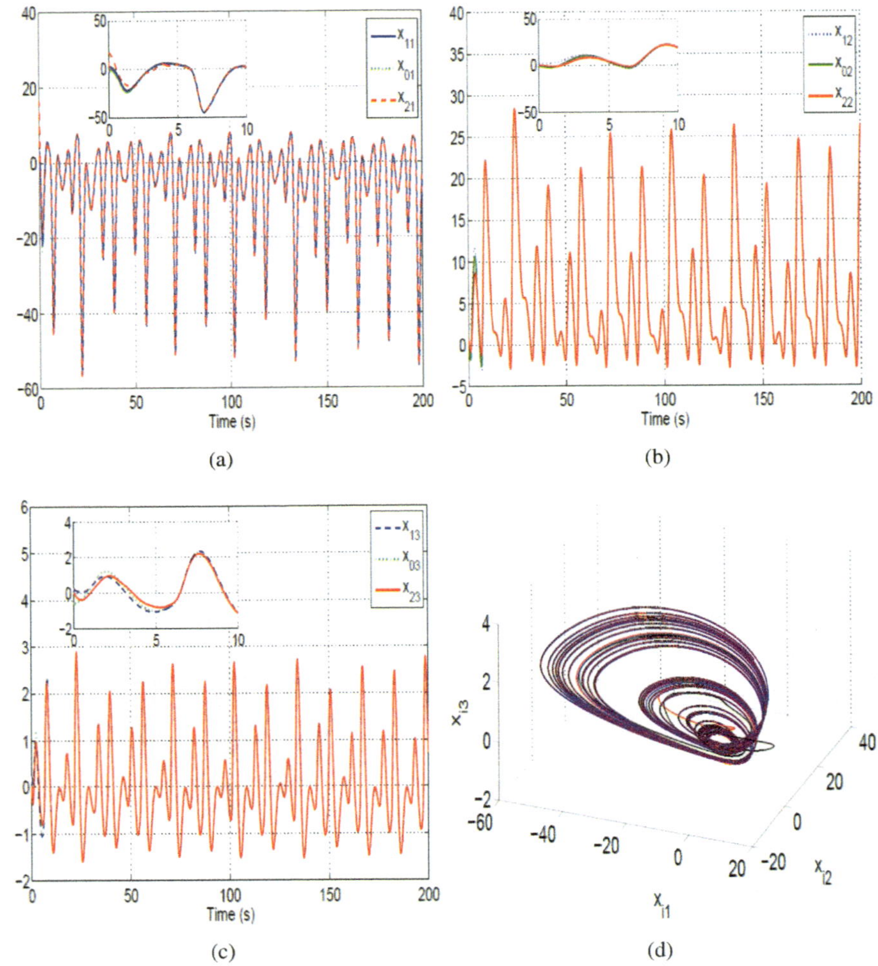

Fig. 4.7 Agreement in original coordinates. **d** Colpitts system attractor, to which the other systems converge. **a–c** Show the state trajectories in the original coordinates

References

1. Ashwin, P., Buescu, J., Stewart, I.: Bubbling of attractors and synchronisation of chaotic oscillators. Phys. Lett. A **193**(2), 126–139 (1994)
2. Boccaletti, S., Kurths, J., Osipov, G., Valladares, D., Zhou, C.: The synchronization of chaotic systems. Phys. Rep. **366**(1–2), 1–101 (2002)
3. Eckmann, J.-P., Ruelle, D.: Ergodic theory of chaos and strange attractors. In: The Theory of Chaotic Attractors, pp. 273–312. Springer (1985)
4. Femat, R., Kocarev, L., Van Gerven, L., Monsivais-Pérez, M.: Towards generalized synchronization of strictly different chaotic systems. Phys. Lett. A **342**(3), 247–255 (2005)
5. Hale, J.K.: Diffusive coupling, dissipation, and synchronization. J. Dyn. Differ. Equ. **9**(1), 1–52 (1997)
6. Heagy, J., Carroll, T., Pecora, L.: Desynchronization by periodic orbits. Phys. Rev. E **52**(2), R1253 (1995)
7. Hramov, A.E., Koronovskii, A.A.: Generalized synchronization: a modified system approach. Phys. Rev. E **71**(6), 067201 (2005)
8. Hunt, B.R., Ott, E., Yorke, J.A.: Differentiable generalized synchronization of chaos. Phys. Rev. E **55**(4), 4029 (1997)
9. Ihle, I., Skjetne, R., Fossen, T.I.: Output feedback control for maneuvering systems using observer backstepping. In: Proceedings of the 2005 IEEE International Symposium on, Mediterrean Conference on Control and Automation Intelligent Control, 2005, pp. 1512–1517. IEEE (2005)
10. Juan, M., Xingyuan, W.: Generalized synchronization via nonlinear control. Chaos: Interdisc. J. Nonlinear Sci. **18**(2), 023108 (2008)
11. Lin, Y., Sontag, E.D., Wang, Y.: A smooth converse Lyapunov theorem for robust stability. SIAM J. Control Optim. **34**(1), 124–160 (1996)
12. Lynch, N. A.: Distributed Algorithms. Elsevier (1996)
13. Martínez-Guerra, R., Pérez-Pinacho, C. A., and Gómez-Cortés, G. C.: Generalized synchronization via the differential primitive element. In Synchronization of Integral and Fractional Order Chaotic Systems, pp. 163–174. Springer (2015)
14. Ott, E., Sommerer, J.C.: Blowout bifurcations: the occurrence of riddled basins and on-off intermittency. Phys. Lett. A **188**(1), 39–47 (1994)
15. Panteley, E., Loria, A.: On practical synchronisation and collective behaviour of networked heterogeneous oscillators. IFAC-PapersOnLine **48**(18), 25–30 (2015)
16. Pecora, L.M., Carroll, T.L.: Synchronization in chaotic systems. Phys. Rev. Lett. **64**(8), 821 (1990)
17. Pecora, L.M., Carroll, T.L.: Driving systems with chaotic signals. Phys. Rev. A **44**(4), 2374 (1991)
18. Pecora, L.M., Carroll, T.L., Heagy, J.F.: Statistics for mathematical properties of maps between time series embeddings. Phys. Rev. E **52**(4), 3420 (1995)
19. Pecora, L. M., Carroll, T. L., Johnson, G. A., Mar, D. J., and Heagy, J. F.: Fundamentals of synchronization in chaotic systems, concepts, and applications. Chaos: An Interdisc. J. Nonlinear Sci. **7**(4), 520–543 (1997)
20. Pikovsky, A.S., Grassberger, P.: Symmetry breaking bifurcation for coupled chaotic attractors. J. Phys. A: Math. Gen. **24**(19), 4587 (1991)
21. Pyragas, K.: Weak and strong synchronization of chaos. Phys. Rev. E **54**(5), R4508 (1996)
22. Qu, Z., Chunyu, J., Wang, J.: Nonlinear cooperative control for consensus of nonlinear and heterogeneous systems. In: 2007 46th IEEE Conference on Decision and Control (2007), pp. 2301–2308. IEEE
23. Rulkov, N.F., Sushchik, M.M., Tsimring, L.S., Abarbanel, H.D.: Generalized synchronization of chaos in directionally coupled chaotic systems. Phys. Rev. E **51**(2), 980 (1995)
24. Saber, R. O., Murray, R. M.: Consensus protocols for networks of dynamic agents

25. Teel, A. R., Praly, L.: A smooth lyapunov function from a class-estimate involving two positive semidefinite functions. In: ESAIM: Control, Optimisation and Calculus of Variations, vol. 5, pp 313–367 (2000)
26. Xu, D., Wang, X., Hong, Y., Jiang, Z.-P.: Global robust distributed output consensus of multi-agent nonlinear systems: an internal model approach. Syst. Control Lett. **87**, 64–69 (2016)

Chapter 5
Multi-synchronization in Heterogeneous Networks

In this chapter, the leader-following consensus problem for networks of strictly different nonlinear systems where is allowed any type of interplay between followers is studied. By considering a directed spanning tree in the network, a dynamic consensus protocol with diffusive coupling signal terms is designed to achieve the leader-following consensus. In this setting, it is revealed that the full closed-loop network can be interpreted as an input-to-state convergent system. Moreover, with the premise that differential-algebraic techniques allow us to completely characterize its synchronization manifold, a stability analysis for this manifold is presented. Finally, the effectiveness of the approach is shown in a numerical example that considers a network of chaotic systems.

5.1 Consensus Problem, Heterogeneous Networks and Interacting Followers

As has been said, consensus and synchronization are closely related problems of multi-agent systems. In [23], this relatlemion is studied by analyzing the complex relationship between the characteristics of the network, i.e., types of systems, coupling and topology involved. Yet, another general description for these problems can be given in terms of the synchronization manifold. If it exists, the behavior of the whole network can be described in terms of this set, whose elements are the variables of interest subject to some constraints due to the characteristics of the network.

A common dynamical behavior is expected within groups of interacting units, i.e., coupled oscillators [1] or multi-agent systems [10]. When this correlated motion occurs, synchronization is achieved for all nodes in the network. It is currently known that it could happen not only for groups of integrator nodes or limit cycle oscillators, but in the case of chaotic systems as well. Thus synchronization manifold can be

R. Martínez-Guerra and J. P. Flores-Flores, *An Approach to Multi-agent Systems as a Generalized Multi-synchronization Problem*, Understanding Complex Systems,
https://doi.org/10.1007/978-3-031-22669-4_5

defined as the stable synchronous solution that results from the interactions of the units in the network.

Compared to the problems of networks of (nearly) identical units [4, 19, 25], the control theory community has tried to solve the problems of synchronization of heterogeneous networks with the appropriate design of dynamic consensus protocols [23]. Recently, a great deal of attention has been given to the study of the latter. Some of those efforts are given in terms of stability in a practical sense [12], where it has been found that the behavior of linearly diffusively coupled units can be synchronized to an emerging motion (synchronization manifold) that depends only on the structure of the network. Other efforts consider the existence of a common virtual dynamical behavior that could be imposed as the case of the exosystem or as the result of the characteristics of the network (endosystem) [23] for all units in the network, such that an appropriate dynamical controller is capable to force all the units to behave in the same manner as the virtual system. The synchronous solution in this sense is given in the form of an internal model principle [22] and an output nonlinear regulator. The way of studying such problems depends on the characterization of the collective behavior between interactive units, i.e., the stability of its synchronization manifold. Hence, the problems just described are very similar to those in consensus literature, it can be seen that the case of the endosystem or emergent dynamics corresponds to a generalization of classical consensus problems and the case of the exosystem to the leader-following consensus problem.

The leader-following consensus problem of multi-agent systems from the last chapter is extended to a case where there is an interplay or interaction between the followers. The synchronous solution to this problem is described in terms of a functional mapping that relates the trajectories of each follower with its leader. Two steps are required: first, find that mapping that also describes the synchronization manifold; and second, verify the stability of that manifold. As stated, the definition of synchronization manifold is a key element to describe the synchronous behavior of the network as a whole. Its stability determines whether a stable synchronous behavior can be achieved [7].

Finding a synchronization manifold depends on the differences between the dynamics of individual nodes in the network. Roughly speaking, for networks of identical units (homogeneous networks) correspond to a trivial synchronization manifold, this is given as the equality of the states; otherwise, for heterogeneous networks, it does not necessarily exist.

In this chapter, it is presented a detailed stability analysis for the full closed-loop network, including the dynamic consensus protocols, where it is revealed that the full closed-loop can be interpreted as an input-to-state convergent system. Moreover, as spanning trees have been proven to be useful within synchronization of networks of interacting linear oscillators [15], here it is proved and extended this condition for nonlinear systems within directed graphs. This means that the existence of directed paths to all followers in the network suffices to reach the leader-following consensus by applying a new distributed dynamic consensus protocol in each follower system.

5.2 Auxiliary Results

Lemma 5.1. *Let* $z, \hat{z} \in \Omega \subset \mathbb{R}^n$ *be vectors in a compact set* Ω *in the Euclidean space and assume the trivial manifold:*

$$\mathscr{M} = \left\{(z, \hat{z}) \in \mathbb{R}^{2n} | z = \hat{z}\right\} \tag{5.1}$$

then

$$\| (z, \hat{z}) \|_{\mathscr{M}} = \frac{1}{\sqrt{2}} \| z - \hat{z} \|_2 \tag{5.2}$$

Proof. From the definition of distance from a point to a set we have:

$$\begin{aligned} \| (z^T, \hat{z}^T)^T \|_{\mathscr{M}} &= \inf_{\left(\xi^t, \hat{\xi}^T\right)^T_{\mathscr{M}}} \left\| \begin{pmatrix} z - \xi \\ \hat{z} - \hat{\xi} \end{pmatrix} \right\|_2 = \left\| \begin{pmatrix} z - \xi \\ \hat{z} - \xi \end{pmatrix} \right\|_2 \\ &= J(\xi) \end{aligned}$$

with $J(\xi) = \sqrt{\|z - \xi\|_2^2 + \|\hat{z} - \xi\|_2^2}$. Note that $J(\xi)$ is a continuous function over a compact set Ω, then $J(\xi)$ attains its infimum at its minimum; hence, $\| \left(z^T, \hat{z}^T\right)^T \|_{\mathscr{M}} = \min_{\xi \in \Omega} J(\xi)$. Any p-norm is convex so we obtain the $argmin J(\xi)$ from first-order sufficient conditions of extrema $\nabla J(\xi) = 0$. Note that $\nabla J(\xi) = \frac{1}{J(\xi)} \left((z - \xi)^T + \left(\hat{z} - \xi\right)^T\right)$. Thus, $argmin J(\xi) = \xi^* = \frac{1}{2}\left(z + \hat{z}\right)$. Finally $\| \cdot \|_{\mathscr{M}}$ is given by

$$\begin{aligned} \| \left(z^T, \hat{z}^T\right)^T \|_{\mathscr{M}} &= \left\| \begin{pmatrix} z - \xi^* \\ \hat{z} - \xi^* \end{pmatrix} \right\|_2 = \frac{1}{2} \inf_{\xi \in \Omega} \left\| \begin{pmatrix} z - \hat{z} \\ -\left(z - \hat{z}\right) \end{pmatrix} \right\|_2 \\ &= \frac{1}{\sqrt{2}} \|z - \hat{z}\|_2 \end{aligned}$$

as desired. □

5.2.1 Convergent Systems

Consider a system

$$\dot{x} = F(t, x), \tag{5.3}$$

where $x \in \mathbb{R}^n$, $t \in \mathbb{R}$ and $F(t, x)$ is locally Lipschitz in x and piecewise continuous in t.

Definition 5.1. (convergent systems [13]). *System* (5.3) *is said to be*

- *convergent if there exists a solution $\bar{x}(t)$ satisfying the following conditions*

(i) *$\bar{x}(t)$ is defined and bounded for all $t \in \mathbb{R}$.*
(ii) *$\bar{x}(t)$ is globally asymptotically stable.*

- *uniformly convergent if it is convergent and $\bar{x}(t)$ is globally uniformly asymptotically stable.*
- *exponentially convergent if it is convergent and $\bar{x}(t)$ is globally exponentially stable.*

Now, consider a time-variant system of the form

$$\dot{x} = F(t, x, \omega) \tag{5.4}$$

where $F : [0, \infty) \times \mathbb{R}^n \times \mathbb{R}^m \to \mathbb{R}^n$ is piecewise continuous in t and locally Lipschitz in x and ω, the input function $\omega : \mathbb{R} \to \mathbb{R}^m$ belongs to the class of piecewise continuous functions with $\|\omega(t)\|_{[t_0,t]} := \sup_{t_0 \le \tau \le t} \|\omega(\tau)\|$ and bounded function of t for all $t \geq 0$.

Definition 5.2. (convergence property with inputs [13]). *System* (5.4) *is said to be uniformly, exponentially convergent if for every piecewise continuous input ω the system $\dot{x} = F(t, x, \omega(t))$ is uniformly, exponentially convergent with limit solution $\bar{x}_\omega(t)$.*

Definition 5.3. input-to-state convergent property [13]). *System* (5.4) *is said to be input-to-state convergent if it is uniformly convergent and for every input ω there exist a $\mathcal{KL}$ function $\beta(r, s)$ and a $\mathcal{K}$ function $\gamma(r)$ such that any solution $x(t)$ of system* (5.4) *corresponding to some bounded piecewise continuous input $\hat{\omega} := \omega + \Delta\omega(t)$ satisfies:*

$$\|x(t) - \bar{x}(t)\| \le \beta\left(\|x(t_0) - \bar{x}(t_0)\|, t - t_0\right) + \gamma\left(\|\Delta\omega\|_{[t_0,t]}\right).$$

Theorem 5.1. *Consider the parallel system*

$$\dot{x}_1 = F_1(t, x_1, \omega, s) \tag{5.5}$$
$$\dot{x}_2 = F_2(t, x_2, \omega, s) \tag{5.6}$$

with $t \in \mathbb{R}$, $x_1 \in \mathbb{R}^{n_1}$, $x_2 \in \mathbb{R}^{n_2}$, $\omega \in \mathbb{R}^{n_3}$ and $s \in \mathbb{R}^{n_4}$. Let the subsystems (5.5) *and* (5.6) *be input-to-state convergent with respect to inputs $\hat{\omega} = \omega(t) + \Delta\omega(t)$ and $\hat{s} = s(t) + \Delta s(t)$, i.e.,*

$$\|\hat{x}_1\| \leq \beta_1\left(\|\hat{x}_1(t_0)\|, t-t_0\right) + \rho_1\left(\sup_{t_0\leq\tau\leq t}\|\Delta\omega(\tau)\|\right) + \gamma_1\left(\sup_{t_0\leq\tau\leq t}\|\Delta s(\tau)\|\right)$$
$$\|\hat{x}_2\| \leq \beta_2\left(\|\hat{x}_2(t_0)\|, t-t_0\right) + \rho_2\left(\sup_{t_0\leq\tau\leq t}\|\Delta\omega(\tau)\|\right) + \gamma_2\left(\sup_{t_0\leq\tau\leq t}\|\Delta s(\tau)\|\right), \tag{5.7}$$

respectively, with $\beta_1, \beta_2 \in \mathcal{KL}$ *and* $\rho_i, \gamma_i \in \mathcal{K}$, $i = 1, 2$. $\|\Delta\omega\| \leq c_1$, $\|\Delta\omega\| \leq c_2$, $c_1, c_2 > 0$, *where* $\hat{x}_1 = x_1 - x_{1.\omega,s}$ *and* $\hat{x}_2 = x_2 - x_{2,\omega,s}$, *where* $x_{1,\omega,s}$ *and* $x_{2,\omega,s}$ *are the steady-state solutions for subsystems* (5.5) *and* (5.6), *respectively. Then the parallel system* (5.5) *and* (5.6) *is an input-to-state convergent system with* ω *and* s *as inputs.*

Proof. Define $\mathcal{X} = \left(\hat{x}_1^T, \hat{x}_2^T\right)^T \in \mathbb{R}^{n_1+n_2}$. For an arbitrary $\mathcal{X}$ it is immediate that $\|X\| \leq \|\hat{x}_1\| + \|\hat{x}_2\|$; thus from (5.7) it follows that

$$\begin{aligned}\|\mathcal{X}\| \leq\ & \beta_1\left(\|\hat{x}_1(t_0)\|, t-t_0\right) + \rho_1\left(\sup_{t_0\leq\tau\leq t}\|\Delta\omega(\tau)\|\right) + \gamma_1\left(\sup_{t_0\leq\tau\leq t}\|\Delta s(\tau)\|\right)\\ & + \beta_2\left(\|\hat{x}_2(t_0)\|, t-t_0\right) + \rho_2\left(\sup_{t_0\leq\tau\leq t}\|\Delta\omega(\tau)\|\right) + \gamma_2\left(\sup_{t_0\leq\tau\leq t}\|\Delta s(\tau)\|\right)\\ \leq\ & \beta_1\left(\|\hat{x}_1(t_0)\|, t-t_0\right) + \beta_2\left(\|\hat{x}_2(t_0)\|, t-t_0\right)\\ & + \max_i\left\{2\rho_i\left(\sup_{t_0\leq\tau\leq t}\|\Delta\omega(\tau)\|\right)\right\} + \max_i\left\{2\gamma_i\left(\sup_{t_0\leq\tau\leq t}\|\Delta s(\tau)\|\right)\right\}\end{aligned}$$

as desired. □

The next result can be considered as a corollary of Property 7 in [13], although a much more direct proof is given here for completeness. This result will be a key element to study the convergence properties of the dynamic consensus protocol.

Theorem 5.2. *Consider the interconnected system*

$$\dot{x}_1(t) = F_1\left(t, x_1, x_2, \omega\right) \tag{5.8}$$
$$\dot{x}_2(t) = F_2\left(t, x_2\right) \tag{5.9}$$

with $t \in \mathbb{R}$, $\omega \in \mathbb{R}^{n_3}$, $x_1 \in \mathbb{R}^{n_1}$ *and* $x_2 \in \mathbb{R}^{n_2}$. *Let the equilibrium point* $x_2 = 0$ *of* x_2*-subsystem be globally uniformly asymptotically stable. Moreover, let subsystem* (5.8) *be input-to-state convergent with respect to inputs* x_2 *and* $\hat{\omega} = \omega(t) + \Delta\omega(t)$:

$$\|\hat{\omega}_1\| \leq \beta_1\left(\|\hat{x}_1(t_0)\|, t-t_0\right) + \rho\left(\sup_{t_0\leq\tau\leq t}\|x_2(\tau)\|\right) + \gamma\left(\sup_{t_0\leq\tau\leq t}\|\Delta\omega(\tau)\|\right) \tag{5.10}$$

with class $\mathcal{KL}$ *function* β_1 *and class* $\mathcal{K}$ *functions* ρ *and* γ, $\|\Delta\omega\| \leq c$, $\hat{x}_1 = x_1 - x_{1\omega}$, *where* $x_{1\omega}$ *is a steady-state solution for convergent system* (5.8). *Then the interconnected system* (5.8) *and* (5.9) *is input-to-state convergent with* ω *as input.*

Proof. Let $t_0 \geq 0$ be the initial time and assume the next inequalities hold globally:

$$\|\hat{x}_1(t)\| \leq \beta_1\left(\|\hat{x}_1(s)\|, t-s\right) + \rho\left(\|x_2(\tau)\|_{[s,t]}\right) + \gamma\left(\|\Delta\omega(\tau)\|_{[s,t]}\right) \tag{5.11}$$

$$\|x_2(t)\| \leq \beta_2\left(\|x_2(s)\|, t-s\right) \tag{5.12}$$

where $t \geq s \geq t_0$, β_1,$\beta_2 \in \mathscr{KL}$ and $\gamma, \rho \in \mathscr{K}$. Let $s = (t+t_0)/2$,

$$\begin{aligned}\|\hat{x}_1(t)\| &\leq \beta_1\left(\|\hat{x}_1((t+t_0)/2)\|, (t-t_0)/2\right)\\ &+ \rho\left(\|x_2(\tau)\|_{[(t+t_0)/2,t]}\right) + \gamma\left(\|\Delta\omega(\tau)\|_{[(t+t_0)/2,t]}\right)\\ &\leq \beta_1\left(\|\hat{x}_1((t+t_0)/2)\|, (t-t_0)/2\right)\\ &+ \rho\left(\|x_2(\tau)\|_{[(t+t_0)/2,t]}\right) + \gamma\left(\|\Delta\omega(\tau)\|_{[t_0,t]}\right). \end{aligned} \tag{5.13}$$

Notice that we can estimate some of the terms in (5.13) from the inequalities in (5.11) and (5.12). First, set $t = (t+t_0)/2$ and $s = t_0$, then estimate $\hat{x}_1((t+t0)/2)$ from (5.11):

$$\begin{aligned}\|\hat{x}_1((t+t_0)/2)\| &\leq \beta_1\left(\|\hat{x}_1(t_0)\|, (t-t_0)/2\right)\\ &+ \rho\left(\|x_2(\tau)\|_{[t_0,(t+t_0)/2]}\right) + \gamma\left(\|\Delta\omega(\tau)\|_{[t_0,(t+t_0)/2]}\right)\\ &\leq \beta_1\left(\|\hat{x}_1(t_0)\|, (t-t_0)/2\right)\\ &+ \rho\left(\|x_2(\tau)\|_{[t_0,(t+t_0)/2]}\right) + \gamma\left(\|\Delta\omega(\tau)\|_{[t_0,t]}\right) \end{aligned} \tag{5.14}$$

and using (5.12), we obtain the following estimates:

$$\begin{aligned}\|x_2(\tau)\|_{[(t+t_0)/2,t]} &\leq \beta_2\left(\|x_2((t+t_0)/2)\|, (t-t_0)/2\right)\\ &\leq \beta_2\left(\|x_2(t_0)\|, (t-t_0)/2\right)\\ \|x_2(\tau)\|_{[t_0,(t+t_0)/2]} &\leq \beta_2\left(\|x_2(t_0)\|, (t-t_0)/2\right)\\ &\leq \beta_2\left(\|x_2(t_0)\|, 0\right) \end{aligned} \tag{5.15}$$

where we have used the fact that x_2 is uniformly asymptotically stable. Thus (5.14) can be rewritten by

$$\begin{aligned}\|\hat{x}_1((t+t_0)/2)\| &\leq \beta_1\left(\|\hat{x}_1(t_0)\|, (t-t_0)/2\right)\\ &\rho\left(\beta_2\left(\|x_2(t_0)\|, 0\right)\right) + \gamma\left(\|\Delta\omega(\tau)\|_{[t_0,t]}\right) \end{aligned} \tag{5.16}$$

From inequalities in (5.13), (5.15) and (5.16), the estimate of $\hat{x}_1(t)$ can be obtained as follows:

$$\begin{aligned}\|\hat{x}_1(t)\| \leq \beta_1 &\left(\beta_1\left(\|\hat{x}_1(t_0)\|, (t-t_0)/2\right)\right.\\ &+ \rho\left(\beta_2\left(\|x_2(t_0)\|, 0\right)\right) + \gamma\left(\|\Delta\omega(\tau)\|_{[t_0,t]}\right), (t-t_0)/2\Big)\\ &+ \rho\left(\beta_2\left(\|x_2(t_0)\|, (t-t_0)/2\right)\right) + \gamma\left(\|\Delta\omega(\tau)\|_{[t_0,t]}\right)\end{aligned}$$

On the other hand, define $\mathscr{X} = (\hat{x}_1^T, x_2^T)^T \in \mathbb{R}^{n_1+n_2}$ and let $s = t_0$ in (5.12), by using the fact that $\|\mathscr{X}\| \leq \|\hat{x}_1(t)\| + \|x_2(t)\|$ and the previous estimate for x_1 then

$$
\begin{aligned}
\|\mathscr{X}\| \leq \beta_1 \big(&\beta_1 \left(\|\hat{x}_1(t_0)\|, (t-t_0)/2\right) + \rho\left(\beta_2\left(\|x_2(t_0)\|, 0\right)\right) \\
&\gamma\left(\|\Delta\omega(\tau)\|_{[t_0,t]}\right), (t-t_0)/2\big) + \rho\left(\beta_2\left(\|x_2(t_0)\|, (t-t_0)/2\right)\right) \\
&+ \gamma\left(\|\Delta\omega(\tau)\|_{[t_0,t]}\right) + \beta_2\left(\|x_2(t_0)\|, (t-t_0)\right)
\end{aligned}
$$

and exploiting the weak triangle inequality [17]:

$$
\begin{aligned}
\|\mathscr{X}\| &\leq \beta_1\left(3\beta_1\left(\|\hat{x}_1(t_0)\|, (t-t_0)/2\right), (t-t_0)/2\right) \\
&+ \beta_1\left(3\rho\left(\beta_2\left(\|x_2(t_0)\|, 0\right)\right), (t-t_0)/2\right) \\
&+ \beta_1\left(3\gamma\left(\|\Delta\omega(\tau)\|_{[t_0,t]}\right), (t-t_0)/2\right) \\
&+ \rho\left(\beta_2\left(\|x_2(t_0)\|, (t-t_0)/2\right)\right) + \gamma\left(\|\Delta\omega(\tau)\|_{[t_0,t]}\right) \\
&+ \beta_2\left(\|x_2(t_0)\|, t-t_0\right) \\
&\leq \beta_1\left(3\beta_1\left(\|\hat{x}_1(t_0)\|, (t-t_0)/2\right), (t-t_0)/2\right) \\
&+ \beta_1\left(3\rho\left(\beta_2\left(\|x_2(t_0)\|, 0\right)\right), (t-t_0)/2\right) \\
&+ \beta_1\left(3\gamma\left(\|\Delta\omega(\tau)\|_{[t_0,t]}\right), 0\right) \\
&+ \rho\left(\beta_2\left(\|x_2(t_0)\|, (t-t_0)/2\right)\right) + \gamma\left(\|\Delta\omega(\tau)\|_{[t_0,t]}\right) \\
&+ \beta_2\left(\|x_2(t_0)\|, t-t_0\right)
\end{aligned} \tag{5.17}
$$

where the last inequality in (5.17) follows from the fact that for $\beta_1 \in \mathscr{KL}$, fixed s_1 and $s_3 \leq s_2$, $\beta_1(s_1, s_2) \leq \beta_1(s_1, s_3)$. Finally next inequality holds

$$
\|\mathscr{X}\| \leq \beta_{\hat{x}_1}\left(\|\hat{x}_1(t_0)\|, t-t_0\right) + \beta_{x_2}\left(\|x_2(t_0)\|, t-t_0\right) + \gamma_{\Delta\omega}\left(\|\Delta\omega\|_\infty\right) \tag{5.18}
$$

with functions $\beta_{\hat{x}_1}, \beta_{x_2} \in \mathscr{KL}$ and $\gamma_{\Delta\omega} \in \mathscr{K}$ given by

$$
\begin{aligned}
\beta_{\hat{x}_1}(r, q) &= \beta_1\left(3\beta_1\left(r, q/2\right), q/2\right) \\
\beta_{x_2}(r, q) &= \beta_1\left(3\rho\left(\beta_2(r, 0)\right), q/2\right) + \rho\left(\beta_2(r, q/2)\right) + \beta_2(r, q) \\
\gamma_{\Delta\omega}(r) &= \gamma(r) + \beta_1(3r, 0)
\end{aligned} \tag{5.19}
$$

as desired. □

Remark 5.1. *The inequality in* (5.10) *uses the fact that second subsystem* (5.9) *is asymptotically stable and zero must be its equilibrium point.*

5.2.2 Graph Theoretical Properties

Theorem 5.3. *[spectral localization [14]] Let $\mathcal{G}_\nu = (\mathcal{V}_\nu, \mathcal{E}_\nu, \mathcal{A}_\nu)$ be a directed graph with the Laplacian $\mathcal{L}_\nu$. Denote the maximum node in-degree of $\mathcal{G}_\nu$ by $\delta(\mathcal{G}_\nu) = \max_i deg_{in}(v_i)$. Then all eigenvalues of $\mathcal{L}_\nu$ are located in the following disc*

$$D(\mathcal{G}_\nu) = \{z \in \mathbb{C} : |z - \delta(\mathcal{G}_\nu)| \leq \delta(\mathcal{G}_\nu)\} \tag{5.20}$$

centered at $z = \delta(\mathcal{G}_\nu) + j0$ in the complex plane. Moreover the real part of the eigenvalues of $\mathcal{L}_\nu$ is nonpositive.

See [11] for proof of this theorem, it follows directly from the Gershgorin Disc Theorem (see Theorem 6.1.1 in [5]). The former theorem is an auxiliary result to prove the next proposition.

Proposition 5.1. *Let $\mathcal{G}_\nu = (\mathcal{V}_\nu, \mathcal{E}_\nu, \mathcal{A}_\nu)$ be a directed graph. Assume $\mathcal{G}_\nu$ has a directed spanning tree with the ν-th node as root, i.e., the Laplacian*

$$\mathcal{L}_\nu = \begin{pmatrix} \mathcal{L}_{\nu-1\times\nu} \\ 0_{1\times\nu} \end{pmatrix} \tag{5.21}$$

where $\mathcal{L}_{\nu-1\times\nu} = (M|b)$, $M = [m_{ij}] \in \mathbb{R}^{N\times N}$ and $b \in \mathbb{R}^N$ with $m_{ij} = -a_{ij}$, $i \neq j$, $m_{ii} = \sum_{j=1,i\neq j}^{N+1} a_{ij}$ and $b = \left(-a_{1(N+1)} \cdots - a_{N(N+1)}\right)^T$, respectively. Then matrix M has all its eigenvalues in the strict right-half of complex plane.

Proof. Because $\mathcal{G}_{N+1}$ is a directed spanning tree with the ν-th node as root, note that all entries of $\mathcal{L}_\nu$ are identically zero in the last row; hence $rank(\mathcal{L}_\nu) = rank(\mathcal{L}_{\nu-1\times\nu}) = \nu - 1$. Furthermore, note that all rows sums of $\mathcal{L}_{\nu-1\times\nu}$ are zero; clearly it follows that the last column of $\mathcal{L}_{\nu-1\times\nu}$ depends on its first $\nu - 1$ columns, i.e., $b = -M\mathbb{1}_{\nu-1}$. Hence $rank(M) = rank(M|b) = \nu - 1$[1]. The latter implies that M is a nonsingular matrix (i.e., $det(M) = \prod_{i=1}^{\nu-1} \lambda_i(M) \neq 0$); equivalently M has no zero eigenvalue. Moreover, as a direct consequence of spectral localization theorem (see Theorem 5.3), we know that all discs of matrix M are on the right-half of complex plane, as no eigenvalue of M is zero then it follows that $Re(\lambda_i(M)) > 0$, $\forall i$. □

5.3 Problem Formulation for a Heterogeneous MAS

Consider the heterogeneous nonlinear multi-agent systems described by the following individual dynamics,

---

[1] The previous statements were given for completeness of exposition and follow ideas from [15].

$$\begin{aligned}\dot{x}_i &= f_i\,(x_i, u_i)\\ \dot{y}_i &= h_i\,(x_i, u_i)\end{aligned} \tag{5.22}$$

where $f_i : \mathbb{R}^n \times \mathbb{R} \to \mathbb{R}^n$ is locally Lipschitz in x_i and u_i; $h : \mathbb{R}^n \times \mathbb{R} \to \mathbb{R}$ is a continuously differentiable function in its arguments; and $x_i = (x_{i1}, \ldots, x_{in}) \in \Omega_i \subset \mathbb{R}^n$, $u_i \in \mathbb{R}$ and $y_i \in \mathbb{R}$ denote the state, input and output of the i-th unit, $i = 1, \ldots, N+1$, respectively.

5.3.1 Heterogeneous MAS and Differential Algebra

Lemma 5.2. ([3]). *Consider an observable system* (5.22) *for i fixed and choose its differential primitive element as:*

$$y_i = \sum_{j=1}^{n} \alpha_j x_{ij} + \sum_{k=1}^{m} \beta_k u_{ik}, \quad \alpha_j, \beta_k \in k < u_i > \tag{5.23}$$

where $u_i = (u_{i1}, \ldots, u_{im})^T$. *System* (5.22) *is Picard-Vessiot if and only if it can be transformable to a GOCF:*

$$\begin{aligned}\dot{\eta}_i &= A_n \eta_i + B_n \left(g_i\,(U_i, \eta_i) + \bar{u}_i\right)\\ \dot{U}_i &= A_\gamma U_i + B_\gamma \bar{u}_i\end{aligned} \tag{5.24}$$

with $\eta_i := \phi_{ui}(x_i) = (y_i, \dot{y}_i, \ldots, y_i^{(n-1)})^T \in \mathbb{R}^n$ *as the control-dependent nonlinear transformation and* $U_i = (u_i, \dot{u}_i, \ldots, u_i^{(\gamma-1)})^T \in \mathbb{R}^\gamma$, $\bar{u}_i \in \mathbb{R}$ *as dynamical control law as a chain of integrators for input* u_i *obtained from the differential primitive element and its* $n-1$ *time derivatives.*

Remark 5.2. *Consider the conditions of Lemma 5.2 and let the differential primitive element be* $y_i = x_{ij} + u_{ij}$. *System* (5.22) *is algebraically observable if and only if* ϕ_{ui}^{-1} *exist. The latter is a direct consequence of the algebraically observable property [9].*

Remark 5.3. *Heterogeneity of nonlinear systems* (5.22) *implies that the vector field associated with each system is distinct, i.e.,* $f_i(x_i, u_i) \neq f_j(x_j, u_j)$. *As a direct consequence* $g_i(U_i, \eta_i) \neq g_j(U_j, \eta_j)$ *in systems* (5.24), $\forall i \neq j$. *This describes the general setting of heterogeneous networks, albeit with some restrictions given by Lemma 5.2.*

Note the following important observations regarding the GOCF. First, assume that system (5.22) can be transformed into its GOCF (5.24). From Lemma 5.2, if the differential primitive element is constrained to be $y_i(t) \equiv 0$, thus $\eta_i(t) = 0$. From (5.24) it immediately implies that

$$
\begin{aligned}
0 &= B_n\left(g_i\left(U_i,0\right)+\bar{u}_i\right)\\
\dot{U}_i &= A_\gamma U_i + \beta_\gamma \bar{u}_i
\end{aligned}
$$

Clearly, input $\bar{u}_i$ is set as

$$
\bar{u}_i = -g_i(U_i,0)
$$

Definition 5.4. *The internal behavior or zero dynamics of system* (5.24) *(respectively for system* (5.22)*) consists of the dynamics*

$$
\dot{U}_i = A_\gamma U_i - B_\gamma g_i\left(U_i,0\right) \tag{5.25}
$$

such that differential primitive element is constrained to be $y_i(t)\equiv 0$ *defined on the maximal interval of existence of solution* η_i *and dynamical control law* U_i.

On the other hand, from Lemma 5.2 it is clear that exact linearization can be achieved under control-dependent transformation given by the differential primitive element.

Example 5.1. *Consider a system in the network as a Rössler system. Thus, its dynamics are given by*

$$
\begin{aligned}
\dot{x}_{11} &= -(x_{12}+x_{13}),\\
\dot{x}_{12} &= x_{11}+a_1x_{12},\\
\dot{x}_{13} &= b_1+x_{13}(x_{11}-c_1),\\
y_1 &= x_{12}+u_1
\end{aligned} \tag{5.26}
$$

where $x_1(0)=(1,2,-5)^T$, $a_1=b_1=0.2$, $c_1=5$. *From its differential primitive element the following coordinate transformation is obtained*

$$
\phi_{u_1}(x_1) := \begin{pmatrix}\eta_{11}\\ \eta_{12}\\ \eta_{13}\end{pmatrix} = \begin{pmatrix} x_{12}+u_{11}\\ x_{11}+a_1x_{12}+u_{12}\\ a_1x_{11}+(a_1^2-1)x_{12}-x_{13}+u_{13}\end{pmatrix},
$$

and its corresponding inverse transformation:

$$
\phi_{u_i}^{-1}(\eta_1) := \begin{pmatrix}x_{11}\\ x_{12}\\ x_{13}\end{pmatrix} = \begin{pmatrix} -a_1(\eta_{11}-u_{11})+\eta_{12}-u_{12}\\ \eta_{11}-u_{11}\\ -(\eta_{11}-u_{11})+a_1(\eta_{12}-u_2)-(\eta_{13}-u_{13})\end{pmatrix},
$$

Then, GOCF for Rössler is given by system (5.24) *with* $\gamma=n$ *and*

$$
\begin{aligned}
g_1(U_1, \eta_1) = & -b_1 - c_1(\eta_{11} - u_{11}) + (a_1c_1 - 1)(\eta_{12} - u_{12}) \\
& + (a_1 - c_1)(\eta_{13} - u_{13}) - a_1(\eta_1 - u_{11})^2 \\
& - a_1(\eta_{12} - u_{12})^2 + (a_1^2 + 1)(\eta_{11} - u_{11})(\eta_{12} - u_{12}) \\
& - a_1(\eta_{11} - u_{11})(\eta_{13} - u_{13}) + (\eta_{12} - u_{12})(\eta_{13} - u_{13})
\end{aligned}
$$

In this case the zero dynamics (5.25) is obtained by making $y_1 = x_{12} + u_{11} = 0$ such that $u_{11} = -x_{12}$, that is,

$$
\begin{aligned}
\dot{U}_i &= A_n U_1 - B_n g_1(U_1, 0) \\
g_1(U_1, 0) &= c_1 u_{11} - (a_1 c_1 - 1)u_{12} - (a_1 - c_1)u_{13} \\
& - a_1 u_{11}^2 - a_1 u_{12}^2 + (a_1^2 + 1)u_{11}u_{12} \\
& - a_1 u_{11}u_{13} + u_{12}u_{13} - b_1
\end{aligned} \tag{5.27}
$$

It is worth mentioning that taking the differential primitive element as $u_{11} = -x_{12}$ gives the internal behavior (5.27) as another GOCF for system (5.26) when $y_1 = 0$.

Remark 5.4. *As seen in the last definition for the type of GOCFs in Lemma 5.2, the zero dynamics of the systems are not involved in the problem of synchronization compared to the case reported in [12]. Here, it is only sufficient to assume that zero dynamics is stable for all systems in the network and some input-to-state convergent property guaranteeing bounded input signals.*

5.3.2 Heterogeneous MAS: Problem Description

Assume that the conditions of Lemma 5.2 are fulfilled. Moreover, consider the spanning tree $\mathscr{G}_{N+1}$ in Fig. 5.1, it models the interaction of the individual units (5.22) and (5.24), where nodes $N+1$ and $\ell = 1, \dots, N$ represent the leader and followers systems, respectively (see. [15], reference model in the linear case). A difference must be made with existing approaches in the sense that consensus protocol for each slave system is dynamic, i.e., it depends on the states of the controller associated with a dynamical system and states of its associated GOCF, which is crucial in the definition of the stable synchronization manifold. All systems (5.22) are said to achieve the leader-following consensus in the sense of the next Definition (illustrated in Fig. 5.2).

Definition 5.5. *The leader-following consensus of multi-agent systems (5.22) is achieved if there exist differential primitive elements that generate the transformations $H_\ell(\cdot) \triangleq \left(\phi_{u_{N+1}}^{-1} \circ \phi_{u_\ell}\right)(\cdot)$, $H_\ell : \Omega_\ell \to \Omega_{N+1}$; an algebraic synchronization manifold $\mathscr{M}_\ell = \{(x_{N+1}, x_l) | x_{N+1} = H_\ell(x_\ell)\}$, a compact set B_ℓ such that $\mathscr{M}_\ell \subset B_\ell \subset \Omega_\ell \times \Omega_{N+1} = D_\ell$ and a dynamic consensus protocol that renders this set the stable attractor of the ℓ-th follower, such that $\lim_{t\to\infty} \|x_{N+1} - H_\ell(x_\ell)\|_2 \to 0$, $\forall (x_{N+1}(t_0), x_l(t_0)) \in B_\ell$, $\forall \ell = 1, \dots, N$.*

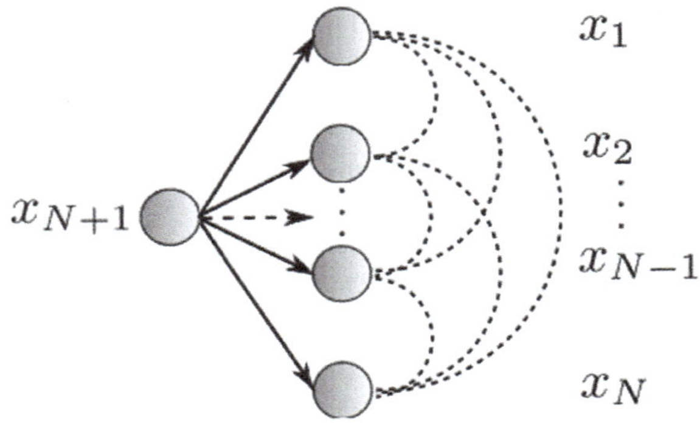

Fig. 5.1 Directed spanning tree $\mathscr{G}_{N+1} = (\mathscr{V}_{N+1}, \mathscr{E}_{N+1}, \mathscr{A}_{N+1})$ in the leader-following consensus problem

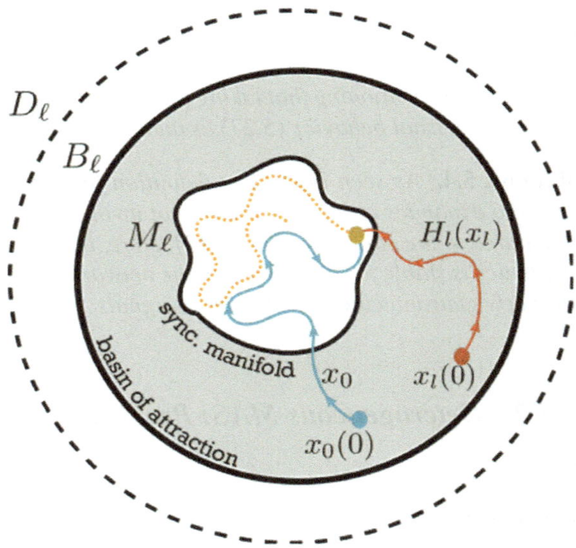

Fig. 5.2 Synchronization manifold of individual units

According to Definition 5.5, an algebraic synchronization manifold for the entire network can be given by

$$\mathscr{M}_x = \{(1_N \otimes x_{N+1}, \bar{x}_\ell) | H_1(x_1) = \cdots = H_N(x_N) = x_{N+1}\} \tag{5.28}$$

where $\bar{x}_\ell = (x_1^T, \ldots, x_N^T)^T \in \mathbb{R}^{Nn}$.

Given $N + 1$ nonidentical systems described by (5.22) and transformable to (5.24), with a fixed topology $\mathscr{G}_{N+1}$ as a directed spanning tree with $N + 1$-th system as root. Assuming complete available information of the state of the systems, find a dynamic consensus protocol such that the algebraic synchronization manifold (5.28) is asymptotically stable, i.e., the leader-following consensus is reached (see Definition 5.5).

Remark 5.5. *This type of consensus problem resembles those of identical linear dynamics such as [16] with a nonobserver-based dynamical controller, in our case for strictly different nonlinear systems with nontrivial synchronization manifold. In [16] there is no explicit leader; however all linear systems should synchronize to an open-loop system (i.e., exosystem [21]) of the form* $\dot{x}_0 = f(t, x_0, 0) = Ax_0$ *where eigenvalues of* A *are in the closed left-half complex plane. Thus exponentially stable synchronization manifold is given by* $\mathscr{M} = \{(x_1, x_2, \ldots, x_N)|x_1 = x_2 = \cdots = x_0\}$. *Here, exponential convergence to the manifold is lost but allows systems to be strictly different. Another approach is given in terms of a nonlinear output regulator [6]; in there the output of the (synchronized) exosystems (local reference nonlinear oscillators) are tracked by their corresponding controlled system, where only those exosystems are coupled. Here the complete state of each coupled follower agrees with the trajectories of the leader within a general stable synchronization manifold, and the coupled systems involved could be chaotic as well.*

5.4 Dynamic Controller for Heterogeneous MAS

In what follows stability is studied. The objective is to impose the dynamics of the leader to all followers in the network such that $\mathscr{M}_x$ is a stable attractor, without loss of generality assume $u_{N+1} = 0$ and that all followers (5.24) are coupled by

$$\begin{aligned}\bar{u}_\ell &= -g_\ell(\eta_\ell, U_\ell) + g_{N+1}(\eta_{N+1}) + c1_n^T(\eta_{N+1} - \eta_\ell) \\ &\quad - c\sum_{j=1}^{N} a_{\ell j} 1_n^T(\eta_\ell - \eta_j) \end{aligned} \tag{5.29}$$

where c is the coupling strength and $a_{\ell j}$ are entries of adjacency matrix

$$\mathscr{A}_{N+1} = \begin{pmatrix} \mathscr{A}_N & 1_N \\ 0_{1\times N} & 0 \end{pmatrix}.$$

Set matrix

$$E = \begin{pmatrix} 0_{(n-1)\times n} \\ 1_n^T \end{pmatrix},$$

thus closed-loop system is given by

$$\begin{aligned}\dot{\eta} &= F_1\left(\eta, g_{N+1}\left(\eta_{N+1}\right)\right) - c\left(\mathscr{L}_{N+1} \otimes E\right)\eta,\\ \dot{\mathscr{U}} &= \left(I_N \otimes A_\gamma\right)\mathscr{U} - c\left(\mathscr{L}_{N\times N+1} \otimes B_r 1_n^T\right)\eta\\ &\quad + 1_N \otimes B_r g_{N+1}\left(\eta_{N+1}\right) - \begin{pmatrix} B_r g_1(\eta_1, U_1)\\ \vdots \\ B_r g_N(\eta_N, U_N)\end{pmatrix}\end{aligned} \tag{5.30}$$

with $\eta = (\eta_1^T, \ldots, \eta_{N+1}^T)^T \in \mathbb{R}^{(N+1)n}$, $F_1(\eta, g_{N+1}(\eta_{N+1})) = (I_{N+1} \otimes A_n)\eta + 1_{N+1} \otimes B_n g_{N+1}(\eta_{N+1})$, $\mathscr{L}_{N+1} = (\mathscr{L}_{N\times N+1}, 0_{1\times N+1})^T$, $\mathscr{L}_{N\times N+1} = (\mathscr{L}_N + I_N - 1_N)$, where the adjacency matrix $\mathscr{A}_N$ and Laplacian matrix $\mathscr{L}_N$ are associated with graph $\mathscr{G}_N = (\mathscr{V}_N, \mathscr{E}_N, \mathscr{A}_N)$ for interactions between followers only. Now, define vector $\bar{\eta}_\ell \triangleq (\eta_1^T, \ldots, \eta_N^T)^T \in \mathbb{R}^{Nn}$ and synchronization error $e \triangleq 1_N \otimes \eta_{N+1} - \bar{\eta}_\ell = (e_1, \ldots, e_N)^T \in \mathbb{R}^{Nn}$, taking the time derivative of e_ℓ, for $1 \le \ell \le N$ we obtain that:

$$\dot{e}_\ell = A_n e_\ell + c\sum_{j=1}^{N+1} a_{ij} E(\eta_\ell - \eta_j) \tag{5.31}$$

by using (5.30) and (5.31) and after some algebraic manipulations

$$\begin{aligned}\dot{e} &= \Xi e,\\ \dot{\mathscr{U}} &= F_2(t, \mathscr{U}, e, \bar{\eta}_\ell),\end{aligned} \tag{5.32}$$

where

$$\begin{aligned}\Xi &= (I_N \otimes (A_n - cE) - c\mathscr{L}_N \otimes E)\\ F_2(t, \mathscr{U}, e, \bar{\eta}_\ell) &= \left(I_N \otimes A_\gamma\right)\mathscr{U} + c\left((\mathscr{L}_N + I_N) \otimes B_r 1_n^T\right)e\\ &\quad + 1_N \otimes B_r g_{N+1}(\eta_{N+1}) - \begin{pmatrix} B_r g_1(\eta_1, U_1)\\ \vdots \\ B_r g_N(\eta_N, U_N)\end{pmatrix}.\end{aligned}$$

Remark 5.6. *In* (5.32), *note that e-subsystem is decoupled from $\mathscr{U}$-subsystem. We can see from the properties of graph $\mathscr{G}_{N+1}$ as being a directed spanning tree with $N+1$-th system as root is a sufficient condition for matrix Ξ to be stable. First, note from Proposition 5.1 that all eigenvalues of $M = \mathscr{L}_N + I_N$ have strict positive real parts. After some algebraic manipulations, it is clear that $\Xi = (I_N \otimes A_n - cM \otimes E)$. By means of Schur form Theorem (see Theorem 2.3.1 in [5]), there exist an unitary matrix $\hat{P} \in \mathbb{R}^{N\times N}$ such that $T = \hat{P}^* M \hat{P}$ is an upper triangular matrix with diagonal entries $T_{ii} = \lambda_i(M)$, by choosing the change of variable $\xi = (\hat{P} \otimes I_n)^{-1} e$*

and taking its time derivative we obtain that[2]

$$\begin{aligned}\dot{\xi}(t) &= (\hat{P} \otimes I_n)^{-1}(I_N \otimes A_n - cM \otimes E)(\hat{P} \otimes I_n)\xi(t) \\ &= (I_N \otimes A_n - cT \otimes E)\xi(t)\end{aligned} \tag{5.33}$$

with

$$T = \begin{pmatrix} \lambda_1(M) & * & \dots & * & * \\ 0 & \lambda_2(M) & \dots & * & * \\ \vdots & 0 & \ddots & \vdots & \vdots \\ 0 & 0 & \dots & \lambda_{N-1}(M) & * \\ 0 & 0 & \dots & 0 & \lambda_N(M) \end{pmatrix}$$

Finally without loss of generality assume real eigenvalues[3] *of* M*; thus the dynamics of* (5.33) *are decoupled in the sense that stability of e-subsystem* (5.32) *is equivalent to stability of the subsystems* $\dot{\xi}_i = (A_n - c\lambda_i(M)E)\xi_i$ *for* $1 \le i \le N$*. Consider* $n = 3$*; hence matrix* Ξ *is stable when* $c = \max_{1 \le i \le N} c_i$, $c_i > 1/\lambda_i(M)$.

Remark 5.7. *The idea of decomposing the dynamics of synchronization error in modes is not new, and it is also the basis for minimum necessary conditions known as Master Stability Function (MSF) approach [2, 18]; however to directly apply MSF approach it is needed that systems to be at least nearly identical [20] and to know beforehand the stable synchronization manifold for the interacting systems [14]. The latter approach can be applied to coupled* η*-subsystem in* (5.30) *but we rather give sufficient conditions on stability of synchronization manifold.*

Theorem 5.4. *Consider a network of* $N + 1$ *heterogeneous multi-agent systems* (5.22)*, as nodes in the directed spanning tree* $\mathcal{G}_{N+1}$*, that can be transformed into a network of systems in their GOCF* (5.24) *coupled by* (5.29)*. If the following assumptions are fulfilled*

(i) choose c *such that* Ξ *is a Hurwitz matrix, i.e., the e-subsystem in* (5.32) *is asymptotically stable.*
(ii) the $\mathcal{U}$*-subsystem in* (5.32) *is input-to-state convergent with respect to* η *and e. Moreover, assume the control input* $\mathcal{U}_R = (\mathcal{U}_{R1}, \mathcal{U}_{R2}, \dots, \mathcal{U}_{RN}) \in \mathbb{R}^{Nn}$, $\mathcal{U}_{R\ell} = \left(u_{R\ell}, \dot{u}_{R\ell}, \dots, u_{R\ell}^{(n-1)}\right) \in \mathbb{R}^n$ *as its steady state.*
(iii) $\phi_{u_{N+1}}^{-1}(\cdot)$ *is continuously differentiable (and uniformly bounded) function with Lipschitz constant* $L > 0$.

Then the closed-loop system (5.32) *is input-to-state convergent and the leader-following consensus is reached, i.e., the algebraic synchronization manifold* $\mathcal{M}_x$ *in* (5.28) *is asymptotically stable.*

---

[2] for any matrices A, B, C, D of appropriate dimension and any constant $k : (A \otimes B)(C \otimes D) = AC \otimes BD$, $(A \otimes B)^{-1} = A^{-1} \otimes B^{-1}$ and $k(A \otimes B) = (kA) \otimes B = A \otimes (kB)$.

[3] eigenvalues of matrix M can be complex, but a similar conclusion will hold under the stability conditions of polynomials with complex coefficients [24].

Proof. First we show that (5.32) is input-to-state convergent; then we show that the synchronization manifold is attractive. To prove the former assume conditions (i)-(ii), then the first result follows from Theorem 5.1. On the other hand, it is clear that if second term of the η-subsystem in (5.32) is zero asymptotically, then $\eta_\ell \to \eta_0$ as $t \to \infty$, $\forall \ell = 1, \dots, N$. To that end assume that (i) $\in c$ such that Ξ is a Hurwitz matrix.

Now, since all nodes behave asymptotically in the same fashion (in transformed coordinates), algebraic synchronization manifold for closed-loop system (5.30) is given by

$$\mathcal{M}_\eta = \{(1_N \otimes \eta_{N+1}, \bar{\eta}_\ell) | \eta_1 = \cdots = \eta_N = \eta_{N+1}\}. \tag{5.34}$$

From Lemma 5.1 the next relation fulfills $\|(1_N \otimes \eta_{N+1} - \bar{\eta}_\ell\|_2 = \sqrt{2}\|(1_N \otimes \eta_{N+1}, \bar{\eta}_\ell)\|_{\mathcal{M}_\eta}$. As Ξ is a Hurwitz matrix we can find matrices $P, Q > 0$ such that $P\Xi + \Xi^T P = -Q$. Let $V = e^T Pe$, by using Rayleigh-Ritz inequality, then

$$\begin{aligned} 2\lambda_{min}(P)\|1_N \otimes \eta_{N+1}, \bar{\eta}_\ell\|^2_{\mathcal{M}_\eta} \le V \le 2\lambda_{max}(P)\|1_N \otimes \eta_{N+1}, \bar{\eta}_\ell\|^2_{\mathcal{M}_\eta}, \\ \dot{V} = e^T(P\Xi + \Xi^T P)e = -e^T Qe \le -\lambda_{min}(Q)\|e\|_2^2. \end{aligned}$$

It is not hard to see from previous inequalities and comparison Lemma [116] that

$$\|(1_N \otimes \eta_{N+1}(t), \bar{\eta}_\ell(t))\|_{\mathcal{M}_\eta} \le \alpha e^{\frac{\beta}{2}t}\|(1_N \otimes \eta_{N+1}(0), \bar{\eta}_\ell(0))\|_{\mathcal{M}_\eta} \tag{5.35}$$

with $\alpha \triangleq \sqrt{\lambda_{max}(P)/\lambda_{min}(P)}$ and $\beta \triangleq \lambda_{min}(Q)/\lambda_{max}(P)$.

On the other hand, from Lemma 5.2 note that $\eta_{N+1} = \phi_{N+1}(x_{N+1})$ and $\bar{\eta}_\ell = (\phi_{u_1}(x_1)^T, \dots, \phi_{u_N}(x_N)^T)^T$, assuming (iii), $\phi^{-1}_{u_{N+1}}(\cdot)$ is a continuously differentiable (and uniformly bounded) function with Lipschitz constant $L > 0$, then

$$\begin{aligned} \left\| 1_N \otimes x_{N+1} - \begin{pmatrix} H_1(x_1) \\ \vdots \\ H_N(x_N) \end{pmatrix} \right\|_2^2 &= \sum_{\ell=1}^{N} \|\phi^{-1}_{u_{N+1}}(\eta_{N+1}) - \phi^{-1}_{u_{N+1}}(\eta_\ell)\|_2^2 \\ &\le L^2 \sum_{\ell=1}^{N} \|\eta_{N+1} - \eta_\ell\|_2^2. \end{aligned} \tag{5.36}$$

It immediately follows from (5.34), (5.35) and above inequality that

$$\left\| 1_N \otimes x_{N+1} - \begin{pmatrix} H_1(x_1) \\ \vdots \\ H_N(x_N) \end{pmatrix} \right\|_2 \le \bar{\alpha} e^{-\frac{\beta}{2}t}\|(1_N \otimes \eta_{N+1}(0), \bar{\eta}_\ell(0))\|_{\mathcal{M}_\eta}$$

with $\bar{\alpha} = \alpha L\sqrt{2}$. Therefore, the leader-following consensus problem for multi-agent systems (5.22) is solved, i.e., $H_\ell(x_\ell) \to x_{N+1}$ as $t \to \infty$, $\forall \ell = 1, \ldots, N$.

Remark 5.8. *First, assumption $i)$ follows from the fact of $\mathscr{G}_{N+1}$ being a directed spanning tree with the $N+1$-th system as root (see Remark 5.6). Assumption $ii)$ is commonly encountered in synchronization literature and it can be seen as the stabilizable condition for the pair (A, B) in linear oscillators [16] or as in our case as part of a separation principle early stated in Theorem 5.2. This second assumption can be checked individually for each system in the network as a direct consequence of Theorem 5.1. Assumption $iii)$ follows from the existence of inverse mapping due to algebraic observability, and this function relates original coordinates from transformed coordinates (see Remark 5.2).*

Remark 5.9. *The assumptions are valid for chaotic systems with $n = 3$ and a single coupling gain c in the network (see Remark 5.6); however this result can be extended to a bigger dimension by letting matrix E to contain different gains in their entries, e.g., $E = (0_{n\times(n-1)}|\kappa^T)^T$, $\kappa = (k_1, k_2, \ldots, k_n) \in \mathbb{R}^{1\times n}$ for $n \geq 3$. With the latter, assumption (i) in Theorem 5.4 implies a computational problem to obtain Hurwitz matrix Ξ as n considerably increases, e.g., $n > 3$ (see [3]).*

Remark 5.10. *The simplicity of this approach is due to differential-algebraic techniques that allow to transform general nonlinear systems (5.22) into its GOCF (5.24) and from the input-dependent transformation, whose time derivatives naturally give the dynamic consensus protocols. Among other things, differential-algebraic techniques let us characterize the algebraic synchronization manifold and to obtain complementary conditions for the approaches based on Master Stability Functions [2].*

5.5 Heterogeneous MAS: Numerical Example

To illustrate Theorem 5.4, consider the numerical example given in [3] with a directed spanning tree $\mathscr{G}_3$ with Rössler, Chua and Colpitts systems as nodes (see Fig. 5.3); where

$$\mathscr{A}_3 = \begin{pmatrix} 0 & 1 & 1 \\ 0 & 0 & 1 \\ 0 & 0 & 0 \end{pmatrix}; \quad \mathscr{L}_3 = \begin{pmatrix} 2 & -1 & -1 \\ 0 & 1 & -1 \\ 0 & 0 & 0 \end{pmatrix}$$

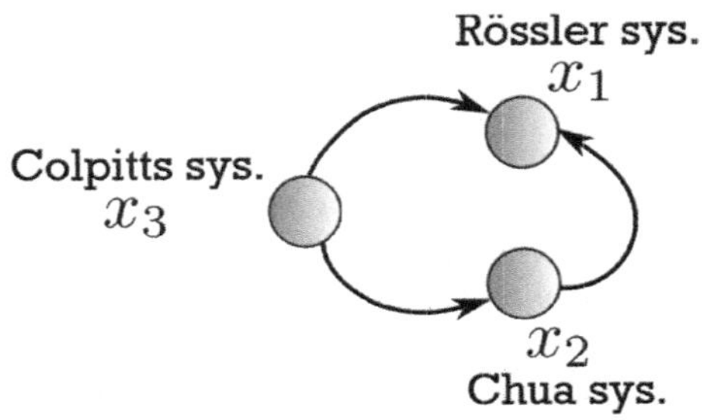

Fig. 5.3 Directed spanning tree in numerical example

Thus, individual dynamics for nodes $i = 1, 2$ are given by

$$\begin{aligned}\dot{x}_{11} &= -(x_{12} + x_{13}),\\ \dot{x}_{12} &= x_{11} + a_1 x_{12},\\ \dot{x}_{13} &= b_1 + x_{13}(x_{11} - c_1),\\ y_1 &= x_{12} + u_1\end{aligned}$$

and

$$\begin{aligned}\dot{x}_{21} &= a_2(x_{22} - x_{21} - \nu_{x_2}),\\ \dot{x}_{22} &= x_{21} - x_{22} + x_{23},\\ \dot{x}_{23} &= -b_2 x_{22}, \nu_{x_2} &&= m_1 x_{21} + 0.5(m_2 - m_1)(|x_{21} + 1| - |x_{21} - 1|),\\ y_2 &= x_{23} + u_2\end{aligned}$$

with coordinate transformation given by

$$\phi_{u_1}(x_1) := \begin{pmatrix}\eta_{11}\\ \eta_{12}\\ \eta_{13}\end{pmatrix} = \begin{pmatrix} x_{12} + u_{11}\\ x_{11} + a_1 x_{12} + u_{12}\\ a_1 x_{11} + (a_1^2 - 1)x_{12} - x_{13} + u_{13}\end{pmatrix},$$

and

$$\phi_{u_2}(x_2) := \begin{pmatrix}\eta_{21}\\ \eta_{22}\\ \eta_{23}\end{pmatrix} = \begin{pmatrix} x_{23} + u_{21}\\ -b_2 x_{22} + u_{22}\\ -b_2(x_{21} - x_{22} + x_{23}) + u_{23}\end{pmatrix},$$

Table 5.1 Parameters for network $\mathscr{G}_3$

i	a_i	b_i	c_i	d_i	m_i
1	0.2	0.2	5	–	−5/7
2	15	25.58	–	–	8/7
3	6.2723	6.2723	0.0797	0.6898	–

Table 5.2 Initial conditions

i	$x_i(0)$
1	$(1.0, 2.0, -5.0)^T$
2	$(0.6, 0.1, 0.6)^T$
3	$(0.6, 0.1, -0.6)^T$

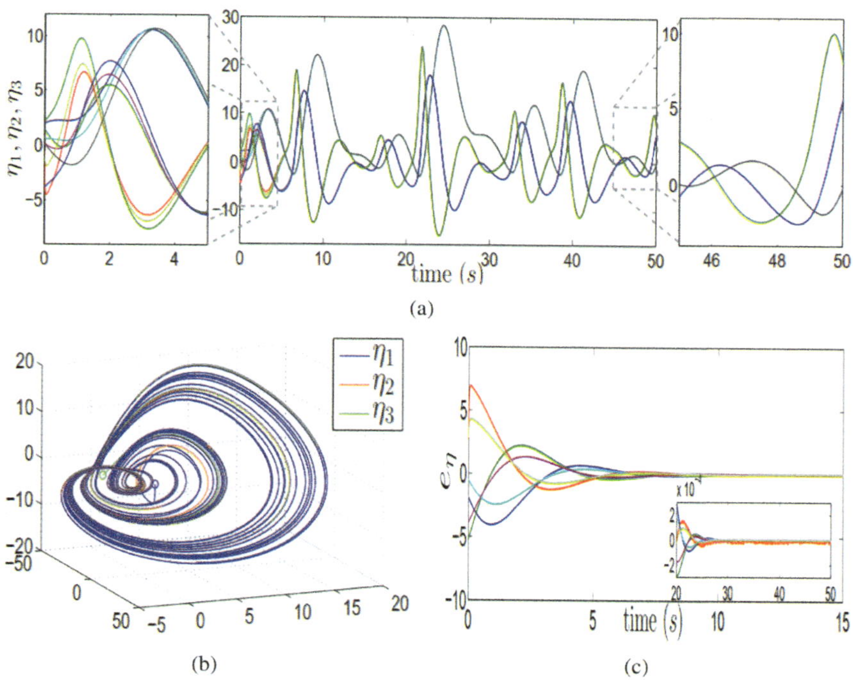

Fig. 5.4 Synchronization of network $\mathscr{G}_3$ with dynamic consensus protocols (5.29) and coupling strength $c = 50$, $e_\eta := e$ and $e_x := 1_2 \otimes x_3 - (H_1(x_1)H_2(x_2))^T$. **a**, **b** Consensus in transformed coordinates, **c** synchronization error

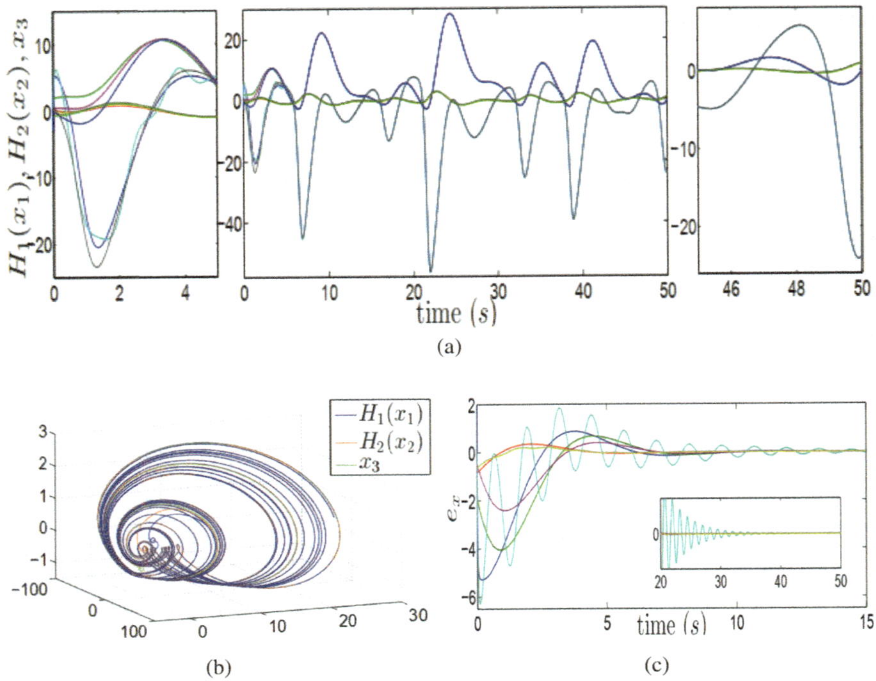

Fig. 5.5 **a**, **b** Leader-following consensus and **c** synchronization error in original coordinates

respectively. And for node $i = 3$

$$\begin{aligned}
\dot{x}_{31} &= -a_3 \exp(-x_{32}) + a_3 x_{33} + a_3,\\
\dot{x}_{32} &= b_3 x_{33},\\
\dot{x}_{33} &= -c_3 x_{31} - c_3 x_{32} - d_3 x_{33},\\
y &= x_{32},
\end{aligned}$$

note that $u_3 = 0$ and inverse transformation $\phi_{u_3}^{-1} \in \mathscr{C}^1$ is given by

$$\phi_{u_3}^{-1} := \begin{pmatrix} x_{31}\\ x_{32}\\ x_{33}\end{pmatrix} = \begin{pmatrix} (-\eta_{33} - d_3\eta_{32} - b_3c_3\eta_{31})/b_3c_3\\ \eta_{31}\\ \eta_{32}/b_3 \end{pmatrix},$$

where $\| [\partial/\partial\eta_3]\, \phi_{u_3}^{-1}(\eta_3)\|_\infty = (b_3c_3 + d_3 + 1) = b_3c_3 =: \bar{L}$, $L \leq \sqrt{3}\bar{L}$. Assume positive constant parameters and initial conditions as in Tables 5.1 and 5.2 ensuring chaotic behavior. And taking into account dynamical control laws (5.29) with coupling strength $c > 1$, such that conditions of Theorem 5.4 are fulfilled.

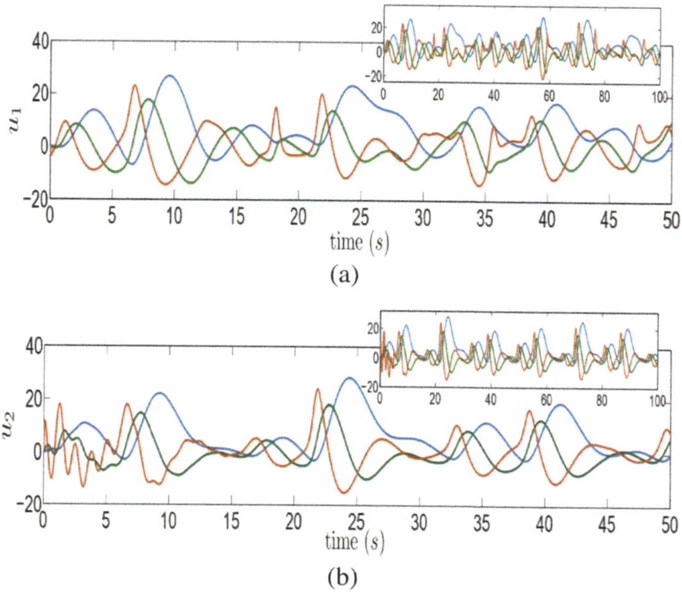

Fig. 5.6 Bounded dynamic consensus protocols (5.29) with coupling strength $c = 50$ for numerical example. **a** Consensus signals of node 1 and **b** of node 2

Choosing $c = 50$, then leader-following consensus is reached in transformed and original coordinates (see Figs. 5.4 and 5.5) where $H_1(x_1(t)) := \phi_{u_3}^{-1} \circ \phi_{u_1}(x_1(t))$ and $H_2(x_2(t)) := \phi_{u_3}^{-1} \circ \phi_{u_2}(x_2(t))$. Notice that the signals of dynamic consensus protocols (5.29) are bounded, and this is shown in Fig. 5.6. Finally an asymptotically stable synchronization algebraic manifold for network $\mathscr{G}_3$ is described by

$$\mathscr{M}_x = \{(1_N \otimes x_3, \bar{x}_\ell) | x_3 = H_1(x_1) = H_2(x_2)\}$$

with $\bar{x}_\ell = (x_1^T, x_2^T)^T$.

References

1. Angeli, D.: A Lyapunov approach to incremental stability properties. IEEE Trans. Autom. Control **47**(3), 410–421 (2002)
2. Arenas, A., Díaz-Guilera, A., Kurths, J., Moreno, Y., Zhou, C.: Synchronization in complex networks. Phys. Rep. **469**(3), 93–153 (2008)
3. Cruz-Ancona, C.D., Martínez-Guerra, R., Pérez-Pinacho, C.A.: Generalized multi-synchronization: a leader-following consensus problem of multi-agent systems. Neurocomputing **233**, 52–60 (2017)
4. Dörfler, F., Bullo, F.: Synchronization in complex networks of phase oscillators: a survey. Automatica **50**(6), 1539–1564 (2014)

5. Horn, R. A., Johnson, C. R.: Matrix Analysis. Cambridge University Press (2012)
6. Isidori, A., Marconi, L., Casadei, G.: Robust output synchronization of a network of heterogeneous nonlinear agents via nonlinear regulation theory. IEEE Trans. Autom. Control **59**(10), 2680–2691 (2014)
7. Kocarev, L., Parlitz, U.: Generalized synchronization, predictability, and equivalence of unidirectionally coupled dynamical systems. Phys. Rev. Lett. **76**(11), 1816 (1996)
8. Liberzon, D.: Switching in Systems and Control. Springer Science & Business Media (2003)
9. Martínez-Guerra, R., Cruz-Ancona, C. D.: Algorithms of Estimation for Nonlinear Systems. Springer (2017)
10. Olfati-Saber, R., Fax, J.A., Murray, R.M.: Consensus and cooperation in networked multi-agent systems. Proc. IEEE **95**(1), 215–233 (2007)
11. Olfati-Saber, R., Murray, R.M.: Consensus problems in networks of agents with switching topology and time-delays. IEEE Trans. Autom. control **49**(9), 1520–1533 (2004)
12. Panteley, E., Loría, A.: Synchronization and dynamic consensus of heterogeneous networked systems. IEEE Trans. Autom. Control **62**(8), 3758–3773 (2017)
13. Pavlov, A., Van De Wouw, N., Nijmeijer, H.: Convergent systems: analysis and synthesis. In: Control and Observer Design for Nonlinear Finite and Infinite Dimensional Systems, pp. 131–146. Springer (2005)
14. Pecora, L.M., Carroll, T.L.: Master stability functions for synchronized coupled systems. Phys. Rev. Lett. **80**(10), 2109 (1998)
15. Ren, W., Beard, R. W.: Distributed Consensus in Multi-vehicle Cooperative Control. Springer (2008)
16. Scardovi, L., Sepulchre, R.: Synchronization in networks of identical linear systems. In: 2008 47th IEEE Conference on Decision and Control, pp. 546–551. IEEE (2008)
17. Sontag, E.D.: Smooth stabilization implies coprime factorization. IEEE Trans. Autom. Control **34**(4), 435–443 (1989)
18. Stilwell, D.J., Bollt, E.M., Roberson, D.G.: Sufficient conditions for fast switching synchronization in time-varying network topologies. SIAM J. Appl. Dyn. Syst. **5**(1), 140–156 (2006)
19. Strogatz, S.H.: From Kuramoto to Crawford: exploring the onset of synchronization in populations of coupled oscillators. Phys. D: Nonlinear Phenom. **143**(1–4), 1–20 (2000)
20. Sun, J., Bollt, E.M., Nishikawa, T.: Master stability functions for coupled nearly identical dynamical systems. EPL (Europhys. Lett.) **85**(6), 60011 (2009)
21. Wieland, P.: From Static to Dynamic Couplings in Consensus and Synchronization Among Identical and Non-identical Systems. Logos Verlag Berlin GmbH (2010)
22. Wieland, P., Allgöwer, F.: An internal model principle for synchronization. In: 2009 IEEE International Conference on Control and Automation, pp. 285–290. IEEE (2009)
23. Wieland, P., Sepulchre, R., Allgöwer, F.: An internal model principle is necessary and sufficient for linear output synchronization. Automatica **47**(5), 1068–1074 (2011)
24. Xie, X.-K.: Stable polynomials with complex coefficients. In: 1985 24th IEEE Conference on Decision and Control, pp. 324–325. IEEE (1985)
25. Zhang, Y., Nishikawa, T., Motter, A.E.: Asymmetry-induced synchronization in oscillator networks. Phys. Rev. E **95**(6), 062215 (2017)

Chapter 6
Synchronization for PDE-Based Systems

The Generalized Synchronization and Generalized Multi-synchronization of nonlinear systems described by partial differential equations (PDE) is studied. The GS is reached utilizing a dynamic controller that is obtained by considering a differential algebraic approach. The key ingredient is the so-called partial differential primitive element. In this context, further extensions of several definitions and results already seen are introduced. Moreover, the stability of the synchronization manifold is analyzed from the point of view of semi-group theory and spectral theory for infinite-dimensional systems. The real importance of this chapter is the study of PDE systems, which appear in most of the real-wold phenomena, and therefore, are crucial for developing real-world applications.

6.1 PDE's and Synchronization

Most real-world phenomena are modeled as *distributed parameter systems* since the state variables of these depend on time and spatial position. The dimensional space of these systems is infinite-dimensional; however, it is very common to approximate them to a finite-dimensional system for application design purposes. Nonetheless, the infinite domain preserves certain interesting properties that are useful for application purposes. For example, an exact representation of the parameter uncertainties or an easy evaluation of the modeling error is possible [13, 16]. In particular, systems described by partial differential equations (*PDE systems*) are crucial for real-world applications. Moreover, from the control theory point of view, several schemes for these systems can be found in the literature. For example, a boundary control is proposed in [4, 17]. A fuzzy control designed with Galerkin's method is presented in [3]. In [14], a convex variation method is used to develop an optimal control. Meanwhile, linear matrix inequalities (LMIs) are used in [25] to find a suitable control.

R. Martínez-Guerra and J. P. Flores-Flores, *An Approach to Multi-agent Systems as a Generalized Multi-synchronization Problem*, Understanding Complex Systems,
https://doi.org/10.1007/978-3-031-22669-4_6

The main disadvantage when working with PDE systems are the required mathematical tools. These tools belong to the fields of differential geometry or calculus of variations and are commonly complex. Moreover, solving problems in these fields requires the solution of an LMI or an adequate Lyapunov functional, which might be hard to solve or find. Thus, the differential algebra approach is even more useful in this context, since it represents a more accessible approach that requires less computational effort [1, 18].

Following the main topic of this book, the synchronization of PDE systems is studied here. This topic is widely unexplored; however, among the related works can be found an output feedback control for reaction-diffusion neural networks in [19]. Itö formula and a Lyapunov–Krasovskii functional method are used in [20] to obtain a controller for stochastic systems. In [26], a distributed state feedback control for parabolic PDE systems is presented. In [22] and [23], the H_∞ method is used to synchronize PDE systems. An impulsive control for spatial-temporal chaotic systems considering the Lyapunov exponent was developed in [12]. In [6], a method for 1D parabolic PDE systems based on the Riez-spectral operator is presented. A feedback controller for second order distributed parameter systems is given in [7] and in [24] is found a nonsingular matrix transformation method which was used to design a boundary control.

As has been said, generalized synchronization is a particular class of synchronization [9, 10]. This chapter presents a definition of generalized synchronization for PDE systems based on the existence of a *partial differential primitive element.* For this, two fundamental aspects are considered: (1) A PDE system can be also seen as a *partial differential field extension* and (2) the solution of a PDE system evolves in an infinite-dimensional Hilbert space. The first aspect leads to the already mentioned GOCF of the involved PDE systems. The second aspect motivates us to use *semi-group* and *spectral theory* for stability analysis of the resultant closed-loop system. It is worth mentioning that, unlike most common analyses, the analysis presented in this chapter allows us to find a very simple criterion for the selection of the control gains.

6.2 GS of PDE Systems by Means of a Dynamical Distributed Control

Let us consider the following nonlinear PDE systems. As usual, let us called the first system as master

$$\Sigma_m := \begin{cases} \dot{\mu}(t,z) &= F_m(\mu, u_m), \quad \mu \in \mathbb{R}^{n_m}, \quad z \in [z_a, z_b] \\ y_m(z,t) &= h_m(\mu) \end{cases}$$

with $\mu(z_a, t) = a(t)$, $\mu(z_b, t) = b(t)$, $\mu(z, 0) = \mu_0$ and $t \in [0, T]$. On the other hand, let the slave system be

$$\Sigma_s := \begin{cases} \dot{\nu}(t,z) &= F_s(\nu, u_s), \quad \nu \in \mathbb{R}^{n_s}, \quad z \in [z_c, z_d] \\ y_s(z,t) &= h_s(\nu) \end{cases}$$

with $\nu(z_c, t) = c(t), \nu(z_d, t) = d(t), \nu(z, 0) = \nu_0$ and $t \in [0, T]$, where $\mathbb{R}^{n_m}$ and $\mathbb{R}^{n_s}$ are the n-dimensional state spaces of master and slave systems, respectively. F_m and F_s are nonlinear functions. The functions h_m and h_s are analytic.

6.2.1 Distributed Dynamical Controller for GS

From a differential algebra point of view, the above systems can also be seen as partial differential field extensions that satisfy certain conditions. This observation is crucial to obtaining the GOCF of these systems. Thus, let us introduce the following definitions.

Definition 6.1. Consider the finite set $\Delta = \{\partial_1, \partial_2, \ldots, \partial_n\}$. A field K provided with the set Δ is a partial differential field, denoted as (K, Δ), if $\partial_1, \ldots, \partial_n$ are derivations $\partial_i : \mathrm{K} \to \mathrm{K}$ such that

(1) $\partial_i(a+b) = \partial_i(a) + \partial_i(b)$
(2) $\partial_i(ab) = \partial_i(a)b + a\partial_i(b)$

For all $a, b \in \mathrm{K}$, $i = 1, 2, \ldots n$.

Definition 6.2. Let $(\mathrm{L}, \Delta_\mathrm{L})$ and $(\mathrm{K}, \Delta_\mathrm{K})$ be partial differential fields with $\Delta_\mathrm{L} = \{\partial_1^\mathrm{L}, \ldots, \partial_n^\mathrm{L}\}$ and $\Delta_\mathrm{K} = \{\partial_1^\mathrm{K}, \ldots, \partial_n^\mathrm{K}\}$. A field extension L/K is a partial differential field extension, denoted as $(\mathrm{L}, \Delta_\mathrm{L})/(\mathrm{K}, \Delta_\mathrm{K})$, if $\mathrm{K} \subset \mathrm{L}$ and if $\partial_i^\mathrm{K}|_\mathrm{L} = \partial_i^\mathrm{L}$, $1 \le i \le n$. This is, all derivation rules over K are restricted by derivation rules over L.

For simplicity, let us refer to a partial differential field (K, Δ) just as K.

Definition 6.3. The element $a \in \mathrm{L}$ is partial differentially algebraic over K if satisfies an algebraic partial differential equation $P(a, \partial a, \ldots, \partial^{(\gamma)}a) = 0$ with coefficients in K and for some $\gamma \in \mathbb{Z}^+$. If every element of L satisfies this condition, the partial differential field extension L/K is said to be partial differentially algebraic.

Definition 6.4. Let $\mathrm{K} < u, y > /\mathrm{K} < u >$ be partial differentially algebraic and $u = \{u_1, u_2, \ldots, u_m\}$ and $y = \{y_1, y_2, \ldots, y_p\}$ be finite sets of partial differential quantities. The partial differential field extension $\mathrm{K} < u, y > /\mathrm{K} < u >$ defines a PDE system with input u and output y.

In definition above, $\mathrm{K} < u >$ denotes the partial differential field generated by K and the finite set of partial differential quantities $u = \{u_1, u_2, \ldots, u_m\}$.

Definition 6.5. A unique element $\bar{y} \in \mathrm{L}$ is a partial differential primitive element of the partial differential field extension L/K if $\bar{y}$ and K differentially generate L, i.e., $\mathrm{K} < \bar{y} >= \mathrm{L}$.

Based on the Definitions 6.3 and 6.4, let us consider the following system:

$$P(\bar{y}, \partial\bar{y}, \ldots, \partial^{(n)}\bar{y}, u, \partial u, \ldots, \partial^{(\gamma)}u) = 0 \tag{6.1}$$

Let n be the minimum integer such that $\partial^{(n)}\bar{y}$ is analytically dependent on $(\bar{y}, \partial\bar{y}, \ldots, \partial^{(n-1)}\bar{y})$. Then, system (6.1) can be locally solved as

$$\partial^{(n)}\bar{y} = \mathscr{L}\left(\bar{y}, \ldots, \partial^{(n-1)}\bar{y}, u, \ldots, \partial^{(\gamma-1)}u\right) + \partial^{(\gamma)}u \tag{6.2}$$

Let us define $\xi_i = \partial^{(i-1)}\bar{y}$, $1 \leq i \leq n$, where n is the so-called *index of algebraic observability*. Then, the system (6.1) can be expressed as the following local representation:

$$\begin{aligned}
\partial\xi_1 &= \xi_2 \\
\partial\xi_2 &= \xi_3 \\
&\vdots \\
\partial\xi_{n-1} &= \xi_n \\
\partial\xi_n &= \mathscr{L}\left(\xi_1, \xi_2, \ldots, \xi_n, u, \ldots, \partial^{(\gamma-1)}u\right) + \partial^{(\gamma)}u \\
y &= \xi_1
\end{aligned} \tag{6.3}$$

Last representation is familiar, i.e., it is the *generalized observability canonical form* of the system (6.1).

Suppose that master and slave systems are transformable into a GOCF.

Definition 6.6. The Systems Σ_m and Σ_s are in a generalized synchronization state if there exists a partial differential primitive element that generates the transformation $H_{ms} : \mathbb{R}^{n_s} \to \mathbb{R}^{n_m}$, with $H_{ms} = \Phi_m^{-1} \circ \Phi_s$, an algebraic manifold $M = \{(\nu, \mu) | \mu = H_{ms}(\nu)\}$ and a compact set $B \subset \mathbb{R}^{n_s} \times \mathbb{R}^{n_m}$, $M \subset B$, such that every trajectory of these systems with initial conditions in B approach to M as $t \to \infty$. This is, if

$$\lim_{t\to\infty} ||H_{ms}(\nu) - \mu|| = 0. \tag{6.4}$$

The transformations $\Phi_m : \mathbb{R}^{n_m} \to \mathbb{R}^{n_c}$ and $\Phi_s : \mathbb{R}^{n_s} \to \mathbb{R}^{n_c}$ from Definition 6.6 are such that $\Phi_m(\mu) = \delta$ and $\Phi_s(\nu) = \eta$. Here, $\mathbb{R}^{n_c}$ denotes a transformed state space that is common for master and slave systems. The vectors δ and η are the state vectors in the transformed coordinates. Hence, the expression (6.4) is equivalent to

$$\lim_{t\to\infty} ||\Phi_m(\mu) - \Phi_s(\nu)|| = 0. \tag{6.5}$$

Now, let us define for each system a partial differential primitive element as a linear combination of the measured variables. For the master system we have

$$\bar{y}_m = \sum_i \alpha_i \mu_i, \quad \alpha_i \in \mathbb{R} < u >$$

and for the slave system

$$\bar{y}_s = \sum_i \beta_i v_i + \sum_j \gamma_j u_j, \quad \beta_i, \gamma_j \in \mathbb{R} < u >$$

These partial differential primitive elements allow us to generate the next coordinate transformations:

$$\Phi_m = \begin{pmatrix} \bar{y}_m \\ \dot{\bar{y}}_m \\ \vdots \\ \bar{y}_m^{(n-1)} \end{pmatrix} = \begin{pmatrix} \delta_1 \\ \delta_2 \\ \vdots \\ \delta_n \end{pmatrix}$$

and

$$\Phi_s = \begin{pmatrix} \bar{y}_s \\ \dot{\bar{y}}_s \\ \vdots \\ \bar{y}_s^{(n-1)} \end{pmatrix} = \begin{pmatrix} \eta_1 \\ \eta_2 \\ \vdots \\ \eta_n \end{pmatrix}$$

One can note that in each transformation, the partial differential primitive element is the first component. Then, the transformation is completed with the time derivatives. Now, the master system can be expressed as

$$\begin{aligned} \dot{\delta}_j &= \delta_{j+1}, \quad 1 \leq j \leq n-1 \\ \dot{\delta}_n &= \mathscr{L}_m(\mu_1, \mu_2, \ldots, \mu_n) \end{aligned} \tag{6.6}$$

while the slave system as

$$\begin{aligned} \dot{\eta}_j &= \eta_{j+1}, \quad 1 \leq j \leq n-1 \\ \dot{\eta}_n &= \mathscr{L}_s(v_1, \ldots, v_n, u, \dot{u}, \ldots, u^{(n-1)}) + u^{(n)} \end{aligned} \tag{6.7}$$

Let the synchronization error be $e(z,t) = \delta - \eta$, $e = (e_1, e_2, \ldots, e_n)^T$. Therefore, from (6.6) and (6.7) we obtain

$$\begin{aligned} \dot{e}_j &= e_{j+1}, \quad 1 \leq j \leq n-1 \\ \dot{e}_n &= \mathscr{L}_m - \mathscr{L}_s - u^{(n)} \end{aligned} \tag{6.8}$$

Then, we define $u^{(n)} = \mathscr{L}_m - \mathscr{L}_s + \sum_{i=1}^{n} \kappa_i e_i$, with $\kappa_i > 0$. Thus, we have the following extended system:

$$\begin{aligned}
\dot{e}_j &= e_{j+1}, \quad 1 \le j \le n-1 \\
\dot{e}_n &= \mathscr{L}_m - \mathscr{L}_s - u_n, \\
\dot{u}_j &= u_{j+1}, \quad 1 \le j \le n-1 \\
\dot{u}_n &= \mathscr{L}_m - \mathscr{L}_s + \sum_{i=1}^{n} \kappa_i e_i
\end{aligned} \tag{6.9}$$

where $u_1 = u, u_2 = \dot{u}, \ldots, u_n = u^{(n-1)}$. The extended system (6.9) defines a *dynamical distributed control* for the generalized synchronization of the systems Σ_m and Σ_s.

6.2.2 Closed-Loop Stability Analysis

The dynamical distributed control given in (6.9) is a finite chain of integrators and leads to a closed-loop system defined by the linear PDE system (6.10). For the stability analysis of this system, we will consider some aspects of semi-group and spectral theory for infinite-dimensional systems. The following definitions and further details can be found in [2, 15].

The linear PDE system

$$\dot{e} = \mathscr{A} e \tag{6.10}$$

where $\mathscr{A}$ is a linear operator on a Hilbert space E has a solution

$$e(t) = S(t)e_0$$

This solution evolves in an infinite-dimensional Hilbert space and is given by the initial conditions $e(0) = e_0$ and the strongly continuous semi-group $S(t)$.

Definition 6.7. [2, 15]. A family of bounded operators $S(t)$, $t \ge 0$, on a Hilbert space E is a strongly continuous semi-group if

(1) $S(0) = I$
(2) $S(t)S(s) = S(t+s)$
(3) $\lim_{s \to t} ||S(s)e - S(t)e|| = 0$

For any $e \in E$ and for all $t, s \ge 0$.

Definition 6.8. [2, 15]. The semi-group $S(t)$ is exponentially stable if there is $\infty > M \ge 1, \alpha > 0$ such that $||S(t)|| \le M \exp(-\alpha t)$ for all $t \ge 0$.

Definition 6.9. [2, 15]. The linear PDE system (6.10) is stable if the linear operator $\mathscr{A}$ generates an exponentially stable semi-group.

Definition 6.10. [2, 15]. A linear operator $\mathscr{A}$ on a Hilbert space E generates a strongly continuous semi-group satisfying $||S(t)|| \le M \exp(-\alpha t)$ if and only if $\mathscr{A}$ is closed, densely defined and $Re(\lambda) < -\alpha$ for every $\lambda \in \sigma(\mathscr{A})$, where $\sigma(\mathscr{A})$ denotes the spectrum of $\mathscr{A}$.

One can note that the exponential stability of the semi-group depends mainly on the parameter α. However, it might be difficult to determine this parameter. Therefore, let us consider the following relation given in [2],

$$r(S(t)) = \exp(\alpha_0 t)$$

where α_0 and $r(S(t))$ are the growth bound and the spectral radius of the semi-group respectively. These are defined as follows.

Definition 6.11. [2]. The growth bound of a strongly continuous semi-group is

$$\alpha_0 = inf\ \{\alpha : ||S(t)|| \leq M(\alpha)\exp(\alpha t),\ M(\alpha) < \infty, \forall t \geq 0\}$$

Definition 6.12. [2]. The spectral radius of the semi-group $S(t)$ at time t is

$$r(S(t)) = \sup\{|\Lambda| : \Lambda \in \sigma(S(t))\}, \quad \sigma(S(t)) \setminus \{0\} = \{\exp(\lambda t) : \lambda \in \sigma(\mathscr{A})\}$$

where $\sigma(S(t))$ is the spectrum of the semi-group.

Theorem 6.1. *The systems Σ_m and Σ_s are in a generalized synchronization state if there exists a dynamical distributed control such that the synchronization error $e = \delta - \eta$ satisfies*

$$\lim_{t\to\infty} \|e(t, z)\| = 0$$

or equivalently

$$\lim_{t\to\infty} \|H_{ms}(\nu) - \mu\| = 0$$

Proof. We can notice that when $Re(\lambda) \leq -\alpha_0 < 0$ for all $\lambda \in \sigma(\mathscr{A})$, $r(S(t)) \to 0$ as $t \to \infty$, i.e., there exists a suitable α such that the strongly continuous semi-group is exponentially stable and therefore the linear PDE system (6.10) is stable.

Observe that in our case, the operator $\mathscr{A}$ of the closed-loop system has the following structure

$$\mathscr{A} = \begin{pmatrix} 0 & 1 & \dots & 0 & 0 \\ \vdots & \vdots & \ddots & \vdots & \vdots \\ 0 & 0 & \dots & 1 & 0 \\ 0 & 0 & \dots & 0 & 1 \\ -\kappa_1 & -\kappa_2 & \dots & -\kappa_{n-1} & -\kappa_n \end{pmatrix}$$

Observe that the matrix $\mathscr{A}$ is a linear bounded operator on a Hilbert space. This operator generates a semi-group defined by the family $\{\exp(\mathscr{A}t)\}$, where

$$\exp(\mathscr{A}t) := \sum_{k=0}^{\infty} \frac{(\mathscr{A}t)^k}{k!}$$

The family $\{\exp(\mathscr{A}t)\}$ is a strongly continuous semi-group since it satisfies $\exp(\mathscr{A}(0)) = I$, $\exp(\mathscr{A}(t+s)) = \exp(\mathscr{A}t)\exp(\mathscr{A}s), \forall t, s \geq 0$. Finally, let us remember that Definition 6.6 is satisfied when $r(\{\exp(\mathscr{A}t)\}) \to 0$ as $t \to \infty$, i.e., when this happens, the linear PDE system (6.10) is stable. □

6.3 GS of PDE Systems: Numerical Results

In what follows, we present two examples to illustrate the GS of PDE systems through a dynamical distributed control. For these examples, we consider the one-dimensional space models of the *Brusselator* system and the *Gray–Scott* system. These systems are reaction-diffusion models and describe the concentration of two substances during a chemical reaction by considering their diffusivity, feed rate, subtraction rate and the interaction between these substances. The reaction-diffusion models correspond to autocatalytic chemical reactions. This kind of process involves a chemical compound that induces and controls a chemical reaction in itself. Complex dynamics, such as multiple equilibrium points, periodic orbits or chaotic behavior, are exhibited by these reactions.

6.3.1 *Brusselator Systems Synchronization*

The Brusselator system was developed by a research team in Brussels in 1960 and describes an oscillating autocatalytic chemical reaction. Its original purpose was to study oscillations and instability in chemical reactions. This system is the simplest reaction-diffusion model that is capable of generating complex spatial patterns. The most simple form of the Brusselator model [5] is governed by the following partial differential equations

$$\begin{aligned}\dot{X}(z,t) &= A - (B+1)X + X^2Y + D_X X_{zz} \\ \dot{Y}(z,t) &= BX - X^2Y + D_Y Y_{zz}\end{aligned} \tag{6.11}$$

The state variables X and Y denote the activator and inhibitor concentrations. The variables D_X and D_Y are diffusion constants. While A and B are the feed rate and the reaction speed respectively. To get different spatial patterns, we can vary these last two constants.

Now, let us consider two Brusselator systems, master and slave. We denote the master system as

$$\begin{aligned}\dot{\mu}_1 &= A_m - (B_m + 1)\mu_1 + \mu_1^2\mu_2 + D_{\mu_1}\mu_{1_{zz}} \\ \dot{\mu}_2 &= B_m\mu_1 - \mu_1^2\mu_2 + D_{\mu_2}\mu_{2_{zz}} \\ y_m &= \mu_1 \end{aligned} \tag{6.12}$$

and the slave system as

$$\begin{aligned}\dot{\nu}_1 &= A_s - (B_s + 1)\nu_1 + \nu_1^2\nu_2 + D_{\nu_1}\nu_{1_{zz}} \\ \dot{\nu}_2 &= B_s\nu_1 - \nu_1^2\nu_2 + D_{\nu_2}\nu_{2_{zz}} \\ y_s &= \nu_1 \end{aligned} \tag{6.13}$$

The behavior of these systems can be observed in Figs. 6.1 and 6.2. Observe that, although these systems have identical parameters, exhibit different spatial patterns

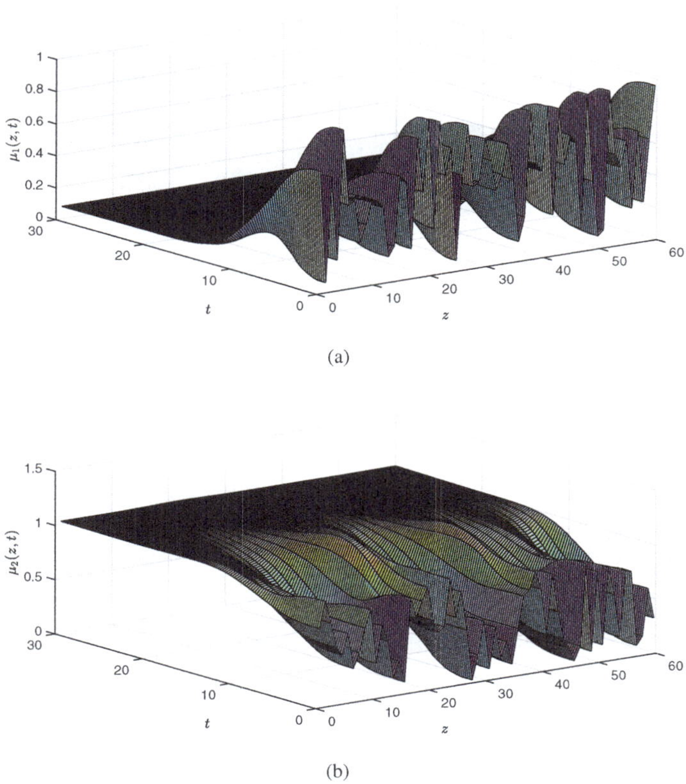

Fig. 6.1 Activator (**a**) and inhibitor (**b**) concentrations of the Brusselator master system with arbitrary initial conditions and feed rate $A_m = 0.09$, reaction speed $B_m = -0.01$ and diffusion constants $D_{\mu_1} = 0.63$, $D_{\mu_2} = 0.8$

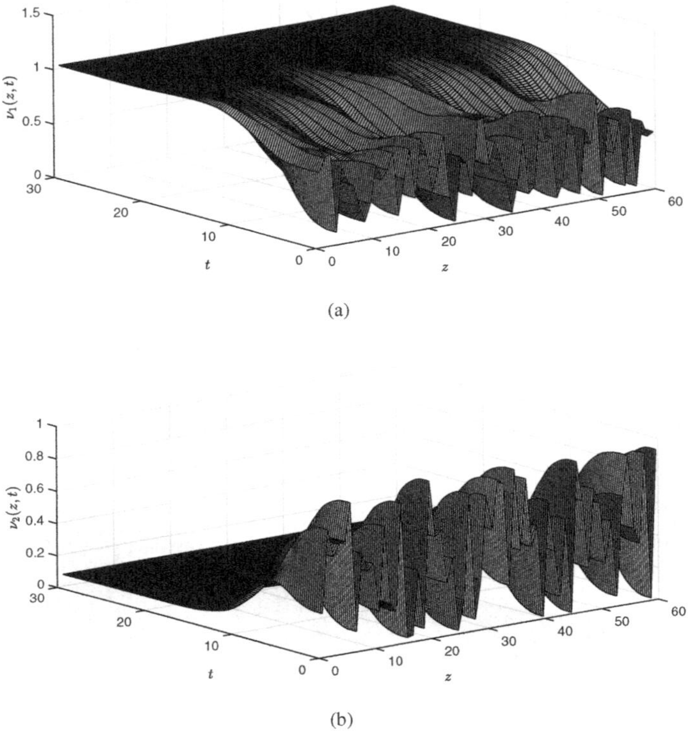

Fig. 6.2 Activator (**a**) and inhibitor (**b**) concentrations of the Brusselator slave system with arbitrary initial conditions and feed rate $A_s = 0.09$, reaction speed $B_s = -0.01$ and diffusion constants $D_{\nu_1} = 0.63$, $D_{\nu_2} = 0.8$

due to the random initial conditions. This is, the Brusselator system has a chaotic behavior.

Now, let $y_m = \delta_1 = \mu_1$ be the partial differential primitive element for the master system. Then, we generate the following transformation

$$\Phi_m(\mu) = \begin{pmatrix} \mu_1 \\ \dot{\mu}_1 \end{pmatrix} = \begin{pmatrix} \delta_1 \\ \delta_2 \end{pmatrix} \tag{6.14}$$

Thus, the system (6.12) can be expressed as

$$\begin{aligned} \dot{\delta}_1 &= \delta_2 \\ \dot{\delta}_2 &= [2\mu_1\mu_2 - (B_m + 1)]\, \dot{\mu}_1 + \mu_1^2 \dot{\mu}_2 + D_{\mu_1} \dot{\mu}_{1_{zz}} \\ &= \mathscr{L}_m(\mu) \end{aligned} \tag{6.15}$$

On the other hand, let $y_s + u_s = \eta_1 = \nu_1 + u_1$ be the partial differential primitive element for system (6.13). Therefore

$$\Phi_s(\nu) = \begin{pmatrix} \nu_1 + u_1 \\ \dot{\nu}_1 + u_2 \end{pmatrix} = \begin{pmatrix} \eta_1 \\ \eta_2 \end{pmatrix} \tag{6.16}$$

such that the slave system is equivalent to

$$\begin{aligned} \dot{\eta}_1 &= \eta_2 \\ \dot{\eta}_2 &= [2\nu_1\nu_2 - (B_s + 1)]\,\dot{\nu}_1 + \nu_1^2\dot{\nu}_2 + D_{\nu_1}\dot{\nu}_{1_{zz}} + u_3 \\ &= \mathscr{L}_s(\nu) + u_3 \end{aligned} \tag{6.17}$$

Then, from (6.15) and (6.17) we have that

$$\begin{aligned} \dot{e}_1 &= e_2 \\ \dot{e}_2 &= \mathscr{L}_m - \mathscr{L}_s - u_3 \end{aligned} \tag{6.18}$$

Therefore, the dynamical distributed control for the generalized synchronization of the systems (6.12) and (6.13) is given by

$$\begin{aligned} \dot{e}_1 &= e_2 \\ \dot{e}_2 &= \mathscr{L}_m - \mathscr{L}_s - u_3 \\ \dot{u}_1 &= u_2 \\ \dot{u}_2 &= u_3 = \mathscr{L}_m - \mathscr{L}_s + \kappa_1 e_1 + \kappa_2 e_2 \end{aligned} \tag{6.19}$$

Thus, the resultant closed-loop system is

$$\dot{e} = \mathscr{A} e = \begin{pmatrix} 0 & 1 \\ -\kappa_1 & -\kappa_2 \end{pmatrix} \begin{pmatrix} e_1 \\ e_2 \end{pmatrix} \tag{6.20}$$

It is not hard to determine that the spectrum of matrix $\mathscr{A}$ is

$$\sigma(\mathscr{A}) = \{\lambda_1, \lambda_2\} = \left\{ -\frac{1}{2}\left(\kappa_2 + \sqrt{-4\kappa_1 + \kappa_2^2}\right), -\frac{1}{2}\left(\kappa_2 - \sqrt{-4\kappa_1 + \kappa_2^2}\right) \right\}$$

In consequence, the spectral radius of the semi-group generated by $\mathscr{A}$ is

$$r(\{\exp(\mathscr{A}t)\}) = \sup\{|\exp(\lambda_1 t)|, |\exp(\lambda_2 t)|\}$$

One can note that for $\kappa_1, \kappa_2 > 0$ we have that $Re(\lambda_1), Re(\lambda_2) < 0$ and therefore $r(\{\exp(\mathscr{A}t)\}) \to 0$ as $t \to \infty$. This is, for $\kappa_1, \kappa_2 > 0$ the system (6.20) is stable and the PDE systems are synchronized.

For simulation purposes, we have discretized the coordinate space by means of the finite difference method. Further details of this method can be found in [21].

The Brusselator systems start from different initial conditions that are set randomly. For these systems, the parameters are identical. The feed rate and the reaction speed are $A_m = A_s = 0.09$ and $B_m = B_s = -0.01$, meanwhile the diffusion constants are $D_{\mu 1} = D_{\nu 1} = 0.63$ and $D_{\mu 2} = D_{\nu 2} = 0.8$. The dynamical distributed control gains are set as $\kappa_1 = 30$ and $\kappa_2 = 10$.

In Figs. 6.3 and 6.4 we observe how activator and inhibitor concentrations of both systems are equal through time in the transformed state space. This affirmation is supported by Fig. 6.5, where we can observe how the synchronization error at each point of the discretized coordinate space tends to zero. Therefore, the Brusselator systems reach a generalized synchronization state. On the other hand, Fig. 6.6 shows the dynamical distributed control signals.

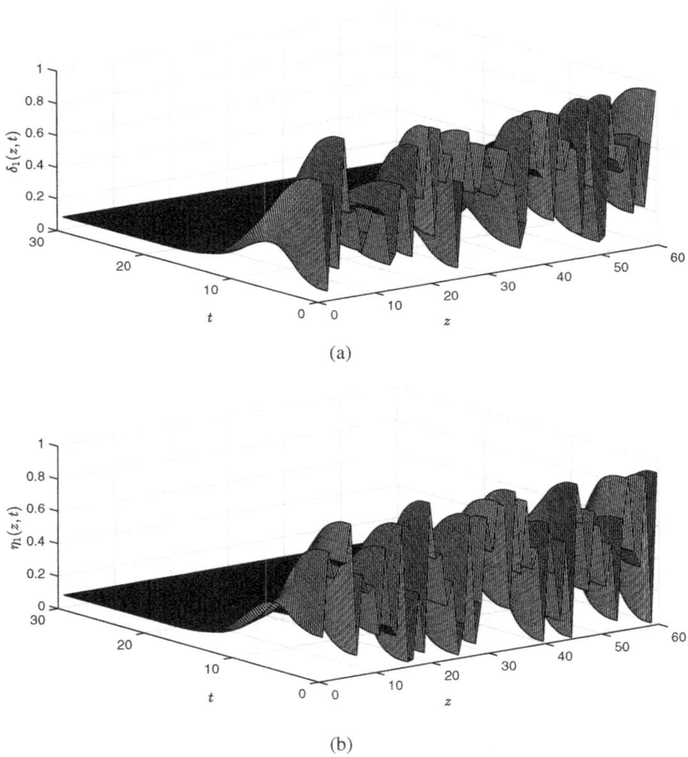

Fig. 6.3 Brusselator systems synchronization: the activator concentration of the master system (**a**) is identical to the slave system's activator concentration (**b**)

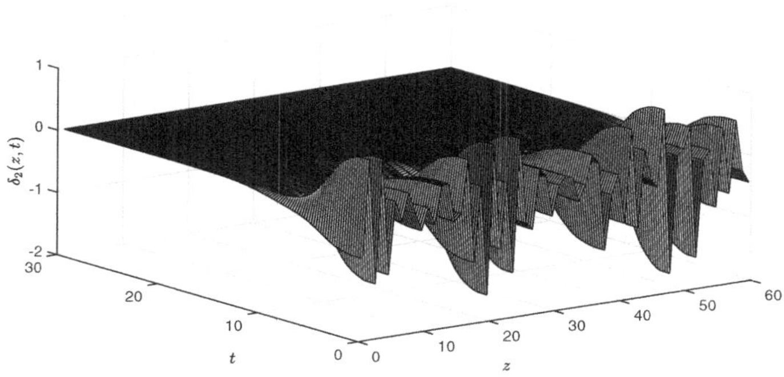

(a)

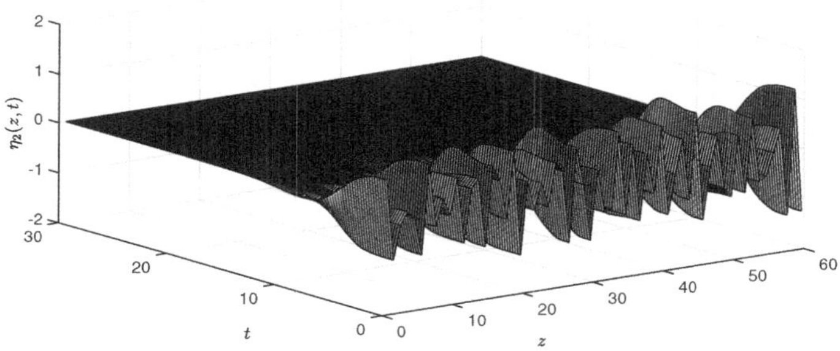

(b)

Fig. 6.4 Brusselator systems synchronization: the inhibitor concentration of the master system (**a**) is identical to the slave system's inhibitor concentration (**b**)

6.3.2 Gray–Scott Systems Synchronization

Let us consider the Gray–Scott system. This system corresponds to two irreversible reactions [11]. The spatio-temporal patterns shown by this system depend on the initial conditions and the system parameters. This system is described by the following PDE system

$$\begin{aligned}\dot{X}(z,t) &= A(1-X) - XY^2 + D_X X_{zz}\\ \dot{Y}(z,t) &= XY^2 - (A+B)Y + D_Y Y_{zz}\end{aligned} \tag{6.21}$$

The state variables X and Y represent the substrate and activator concentrations. Variable A represents the feed rate, and B is the reaction speed. Meanwhile, D_X and D_Y are diffusion constants.

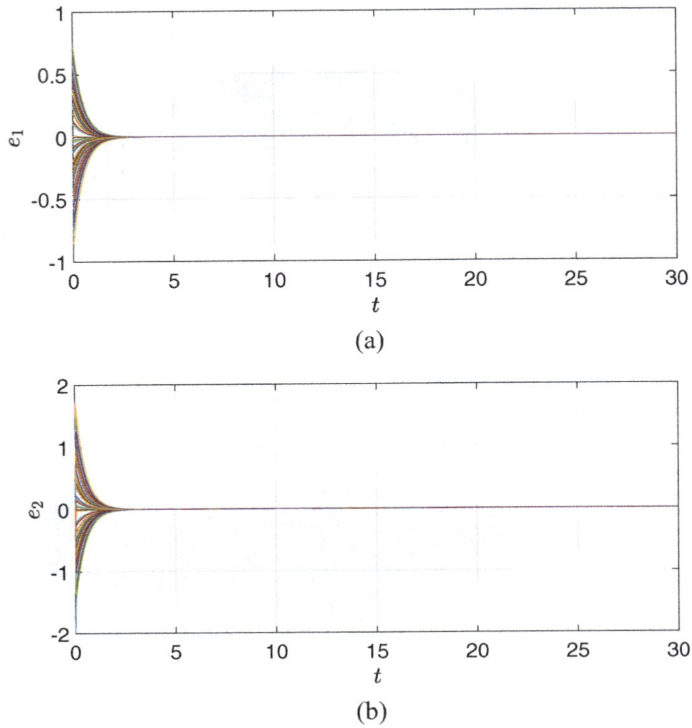

Fig. 6.5 Brusselator systems synchronization: synchronization error for activator (**a**) and inhibitor (**b**) concentrations at each point of the discretized space

Let us consider two Gray–Scott systems. The master system is described by

$$
\begin{aligned}
\dot{\mu}_1 &= A_m(1-\mu_1) - \mu_1\mu_2^2 + D_{\mu_1}\mu_{1_{zz}} \\
\dot{\mu}_2 &= \mu_1\mu_2^2 - (A_m + B_m)\mu_2 + D_{\mu_2}\mu_{2_{zz}} \\
y_m &= \mu_1
\end{aligned} \tag{6.22}
$$

while the Gray–Scott slave system is

$$
\begin{aligned}
\dot{\nu}_1 &= A_s(1-\nu_1) - \nu_1\nu_2^2 + D_{\nu_1}\nu_{1_{zz}} \\
\dot{\nu}_2 &= \nu_1\nu_2^2 - (A_s + B_s)\nu_2 + D_{\nu_2}\nu_{2_{zz}} \\
y_s &= \nu_1
\end{aligned} \tag{6.23}
$$

We can observe the behavior of these PDE systems in Figs. 6.7 and 6.8. Notice that the systems start from random initial conditions, and even more, these have different parameters. In consequence, they exhibit different spatial patterns [8].

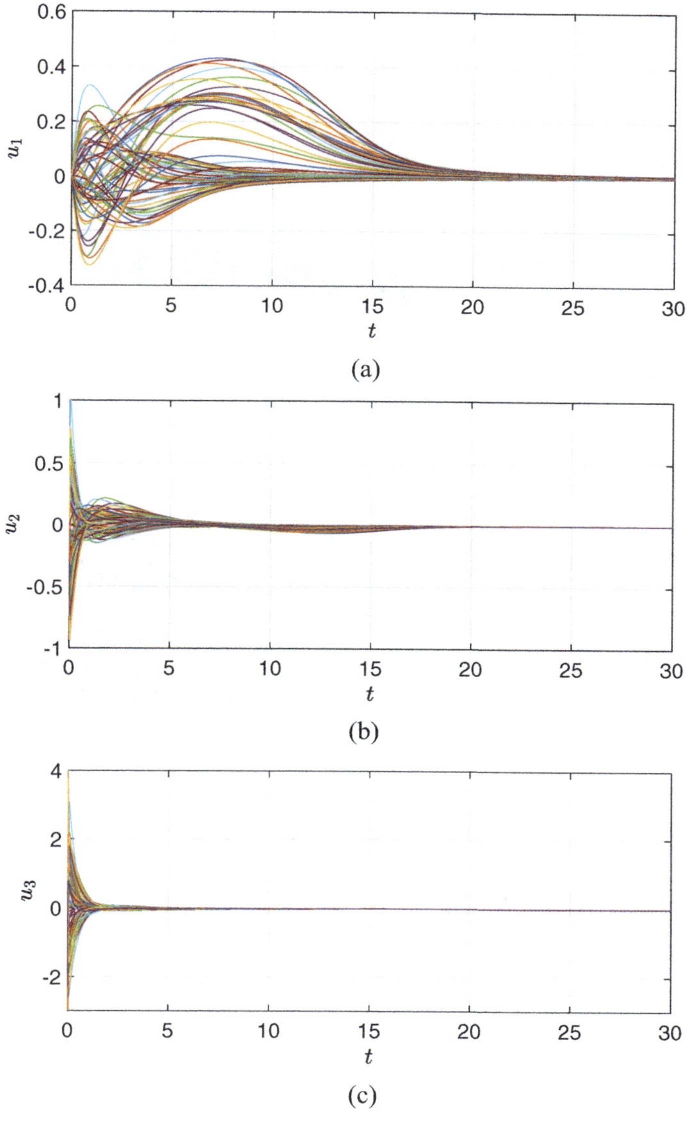

Fig. 6.6 **a** Control signals for the synchronization of the Brusselator systems at each point of the discretized space and its **b** first and **c** second integrals

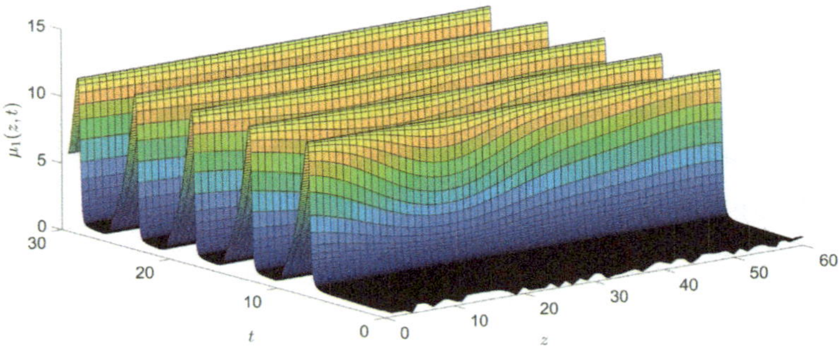

(a)

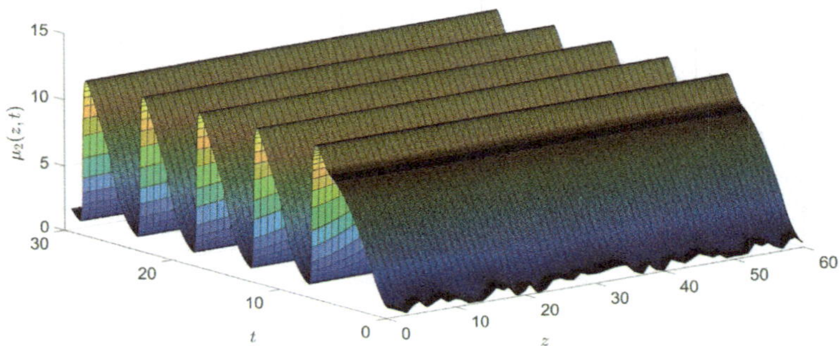

(b)

Fig. 6.7 Substrate (**a**) and activator (**b**) concentrations for the Gray–Scott master system with random initial conditions and feed rate $A_m = 2.5$, reaction speed $B_m = 9$ and diffusion constants $D_{\mu_1} = 7$, $D_{\mu_2} = 10$

Similar to the first example, we choose $y_m = \mu_1$ and $y_s + u_s = \nu_1 + u_1$ as partial differential primitive elements for master and slave system respectively, such that these can be expressed as

$$\begin{aligned}\dot{\delta}_1 &= \delta_2 \\ \dot{\delta}_2 &= D_{\mu_1}\dot{\mu}_{1_{zz}} - \left(\mu_2^2 + A_m\right)\dot{\mu}_1 - 2\mu_1\mu_2\dot{\mu}_2 \\ &= \Psi_m(\mu)\end{aligned} \tag{6.24}$$

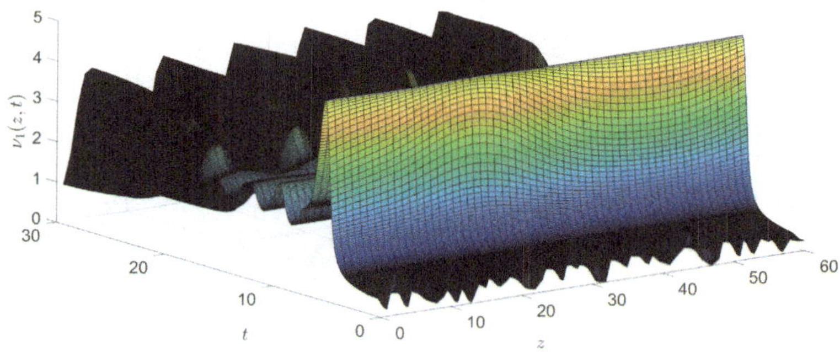

(a)

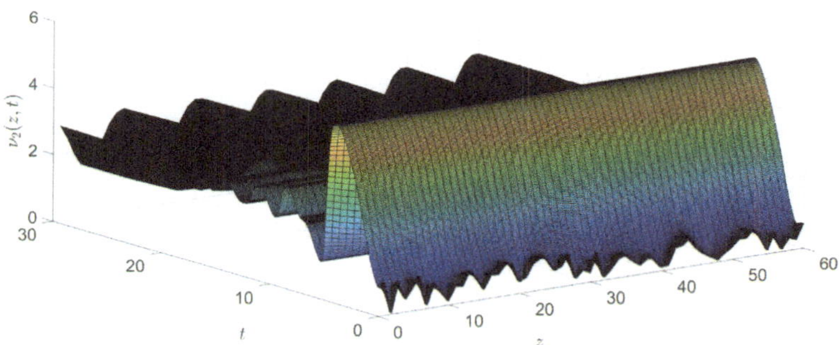

(b)

Fig. 6.8 Substrate (**a**) and activator (**b**) concentrations for the Gray–Scott slave system with random initial conditions and feed rate $A_s = 2$, reaction speed $B_s = 4.8$ and diffusion constants $D_{\nu_1} = 2$, $D_{\nu_2} = 10$

and as

$$\begin{aligned}
\dot{\eta}_1 &= \eta_2 \\
\dot{\eta}_2 &= D_{\nu_1} \dot{\nu}_{1_{zz}} - \left(\nu_2^2 + A_s\right) \dot{\nu}_1 - 2\nu_1 \nu_2 \dot{\nu}_2 + u_3 \\
&= \Psi_s(\nu) + u_3
\end{aligned} \tag{6.25}$$

Then, from (6.24) and (6.25) we have that dynamical distributed control for the synchronization of these systems is

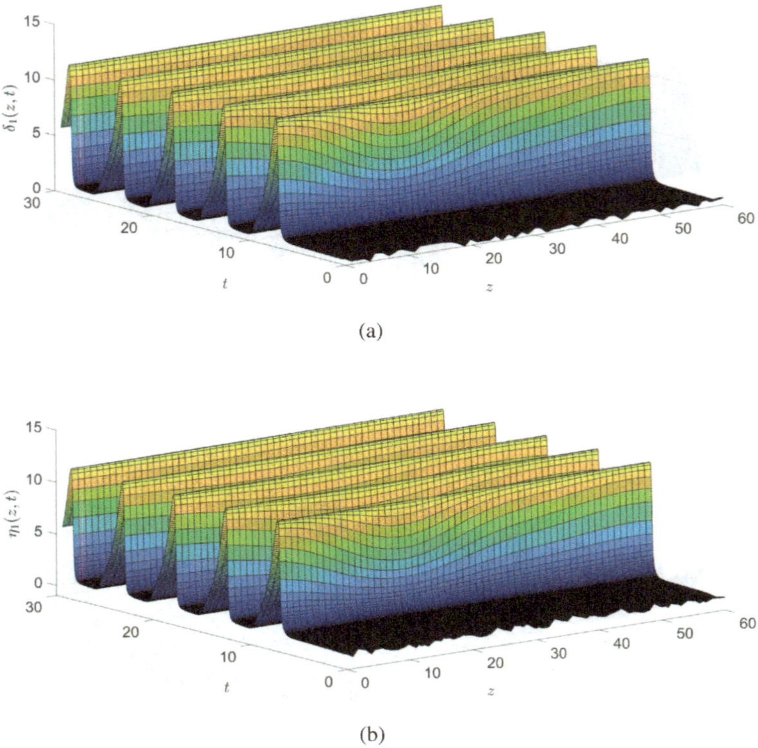

Fig. 6.9 Gray–Scott systems synchronization: the substrate concentration of the master system (**a**) is identical to the slave system's substrate concentration (**b**)

$$\begin{aligned}\dot{e}_1 &= e_2\\ \dot{e}_2 &= \Psi_m - \Psi_s - u_3\\ \dot{u}_1 &= u_2\\ \dot{u}_2 &= u_3 = \Psi_m - \Psi_s + \kappa_1 e_1 + \kappa_2 e_2\end{aligned} \tag{6.26}$$

Hence, the resultant closed-loop system is $\dot{e} = \mathscr{A} e$, where

$$\mathscr{A} = \begin{pmatrix} 0 & 1 \\ -\kappa_1 & -\kappa_2 \end{pmatrix}$$

Finally, just as in the example above, master and slave systems reach a state of generalized synchronization when $\kappa_1, \kappa_2 > 0$.

In what follows, we present the numerical results. As in the first example, we use the finite difference method to discretize the coordinate space. The Gray–Scott systems start from random initial conditions. The feed rate and reaction speed of

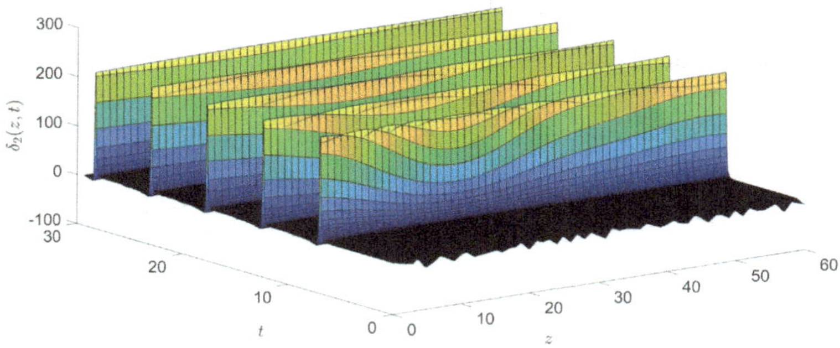

(a)

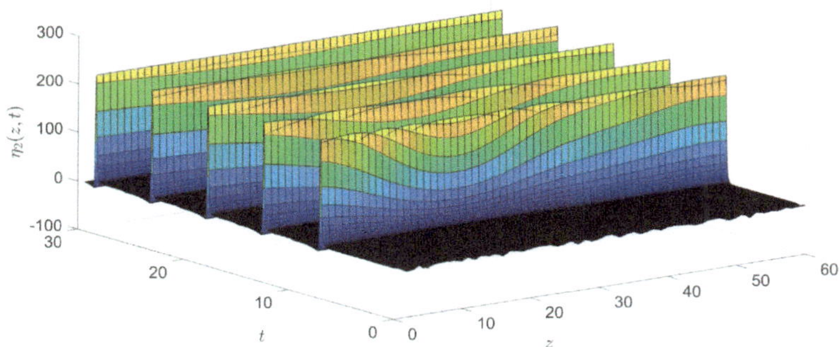

(b)

Fig. 6.10 Gray–Scott systems synchronization: the activator concentration of the master system (**a**) is identical to the slave system's activator concentration (**b**)

the master system are $A_m = 2.5$ and $B_m = 9$, the diffusion constants are $D_{\mu 1} = 7$ and $D_{\mu 2} = 10$. For slave system we have a feed rate $A_s = 2$ and a reaction speed $B_s = 4.8$ with diffusion constants $D_{\nu 1} = 2$ and $D_{\nu 2} = 10$. The dynamical distributed control gains are set as $\kappa_1 = 100$ and $\kappa_2 = 50$.

In Figs. 6.9 and 6.10 we observe the behavior of the systems in the transformed coordinate space. One can note that the concentrations are identical in both systems. In Fig. 6.11 we corroborate that a generalized synchronization state has been reached since the synchronization error goes to zero. Finally, in Fig. 6.12 we observe the signals generated by the dynamical distributed control.

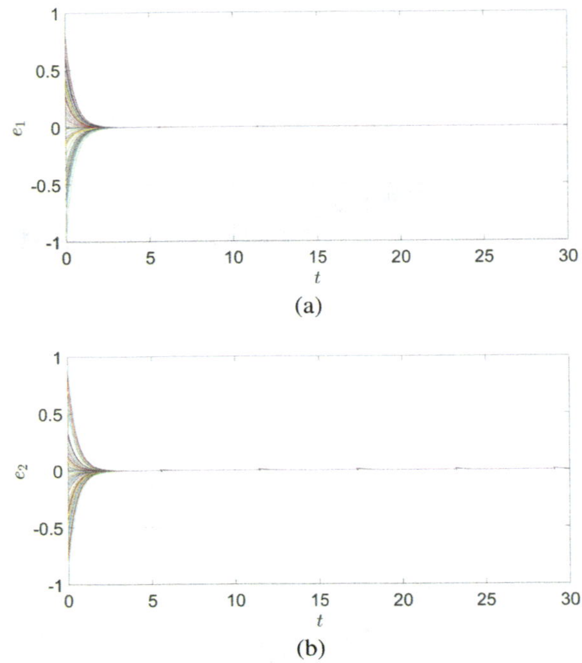

Fig. 6.11 Gray–Scott systems synchronization: synchronization error for substrate (**a**) and activator (**b**) concentrations at each point of the discretized space

6.4 Multi-synchronization of PDE Systems

The synchronization of multiple PDE-based systems also can be reduced to a generalized multi-synchronization (GMS). For this purpose, the involved systems are grouped in two families. The first family, which contains the systems to be followed, is called master family. On the other hand, the family of followers will be named as slave family. Then, a family of dynamical distributed controllers is designed to reach a synchronous state for all the systems.

Let us consider the following family of master systems (master family)

$$\Sigma_m := \begin{cases} \dot{V}_\rho(x,t) &= R_{m_\rho}[x]V_\rho(x,t) + Q_{m_\rho}(x,t,u_{m_\rho}), \\ y_{m_\rho}(x,t) &= h_{m_\rho}(V_\rho, u_{m_\rho}), \end{cases}$$

where $V_\rho \in \mathbb{R}^{n_{m_\rho}}$, $x \in [0, L]$. And a second family of slave systems (slave family)

$$\Sigma_s := \begin{cases} \dot{W}_\sigma(x,t) = R_{s_\sigma}[x]W_\sigma(x,t) + Q_{s_\sigma}(x,t,u_{s_\sigma}), \\ y_{s_\sigma}(x,t) \;= h_{s_\sigma}(W_\sigma, u_{s_\sigma}), \end{cases}$$

where $W_\sigma \in \mathbb{R}^{n_{s_\sigma}}$, $x \in [0, L]$. Indexes ρ and σ are such that $1 \le \sigma \le p-1$, $1 \le \rho \le p - \sigma$. These two conditions ensure that each system in the master family can be

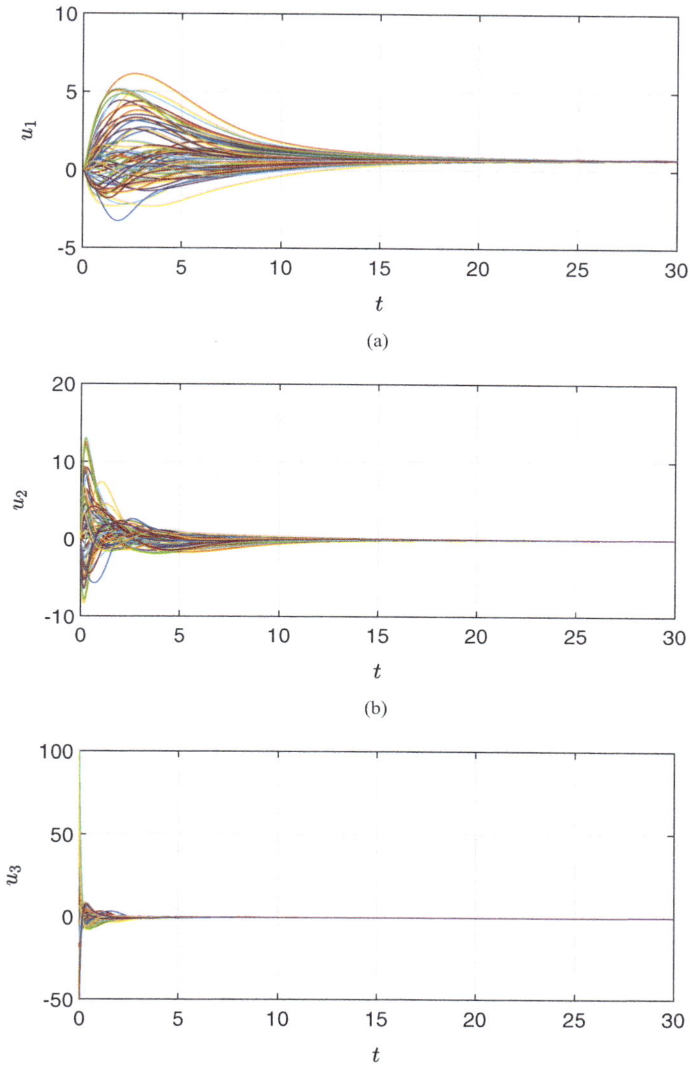

Fig. 6.12 **a** Control signals for the synchronization of the the Gray–Scott systems at each point of the discretized space and its **b** first and **c** second integrals

associated with multiple systems from the slave family; nonetheless, every system in the slave family is associated with a unique system in the master family. This scenario is visualized in Fig. 6.13.

Definition 6.13. Let $\bar{V} = (V_1, V_2, \ldots, V_{n_m}) \in \mathbb{R}^{n_m}$ and $\bar{W} = (W_1, W_2, \ldots, W_{n_s}) \in \mathbb{R}^{n_s}$ be the master and slave state vector families, respectively. The families Σ_m and

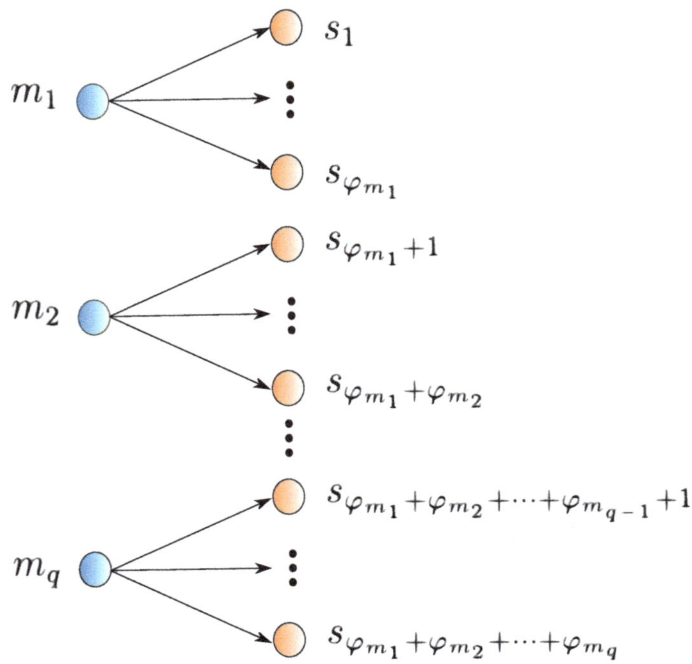

Fig. 6.13 Interaction between systems of master family and salve family. Here the suffixes φ_{m_i} denote the number of slave systems that interact with the master system m_i

Σ_s are said to be in a state of generalized multi-synchronization (GMS) if there exists a family of changes of variable that generates a family of transformations $H : \mathbb{R}^{n_s} \to \mathbb{R}^{n_m}$, with $H = \Psi_m^{-1} \circ \Psi_s$, an algebraic manifold $\mathscr{M} = \{(\bar{W}, \bar{V}) | \bar{V} = H(\bar{W})\}$ and a compact set $B \subset \mathbb{R}^{n_s} \times \mathbb{R}^{n_m}$, $\mathscr{M} \subset B$, such that every trajectory of these families with initial conditions in B approach to $\mathscr{M}$ as $t \to \infty$. Therefore, master and slave families are in a GMS state if

$$\lim_{t\to\infty} ||H(\bar{W}) - \bar{V}|| = 0. \tag{6.27}$$

As the families of transformations Ψ_m and Ψ_s take both state vector families into a common dimensional space, this is, $\Psi_m : \mathbb{R}^{n_m} \to \mathbb{R}^{n_\delta}$ and $\Psi_s : \mathbb{R}^{n_s} \to \mathbb{R}^{n_\delta}$, we can say that condition (6.27) is equivalent to

$$\lim_{t\to\infty} ||\Psi_m(\bar{V}) - \Psi_s(\bar{W})|| = 0. \tag{6.28}$$

Theorem 6.2. *If each system of the families Σ_m and Σ_s is transformable to a GOCF, then there exists a family of dynamical distributed controllers such that*

$\lim_{t\to\infty} ||\Psi_m(\bar{V}) - \Psi_s(\bar{W})|| = 0$, *i.e., the families of integer order PDE-based systems Σ_m and Σ_s are in a GMS state.*

Proof. Let us consider that the master family has "q" systems and the slave family has "r" systems. Then, we define the following family of changes of variable for master family

$$y_{m_j} = \sum_{i=n_m-n_{m_j}+1}^{n_m} \beta_i x_{i,m_j} = \delta_l^{n_{m_j}}, \quad 1 \le j \le q, \quad 1 \le i \le \sum_{1\le j\le q} n_{m_j} = n_m,$$
$$l = 1, n_{m_1}+1, \ldots, n_{m_1}+n_{m_2}+\cdots+n_{m_{q-1}}+1$$

and for slave family,

$$y_{s_j} = \sum_{i=n_s-n_{s_j}+1}^{n_s} \beta_i x_{i,m_j} + \sum_k \zeta_k u_{k,s_j} = \delta_l^{n_{s_j}}, \quad 1 \le j \le r,$$
$$1 \le i \le \sum_{1\le j\le r} n_{s_j} = n_s,$$
$$l = 1, n_{s_1}+1, \ldots, n_{s_1}+n_{s_2}+\cdots+n_{s_{r-1}}+1$$

where β_i and ζ_k are constants. Therefore, we can generate the next families of transformations for master family and slave family, respectively:

$$\Psi_m = \begin{pmatrix} y_{m_1} \\ \dot{y}_{m_1} \\ \vdots \\ y_{m_1}^{(n_{m_1}-1)} \\ y_{m_2} \\ \dot{y}_{m_2} \\ \vdots \\ y_{m_2}^{(n_{m_2}-1)} \\ \vdots \\ y_{m_q} \\ \dot{y}_{m_q} \\ \vdots \\ y_{m_q}^{(n_{m_q}-1)} \end{pmatrix} = \begin{pmatrix} \delta_1^{n_{m_1}} \\ \delta_2^{n_{m_1}} \\ \vdots \\ \delta_{n_{m_1}}^{n_{m_1}} \\ \delta_{n_{m_1}+1}^{n_{m_2}} \\ \delta_{n_{m_1}+2}^{n_{m_2}} \\ \vdots \\ \delta_{n_{m_1}+n_{m_2}}^{n_{m_2}} \\ \vdots \\ \delta_{n_{m_1}+n_{m_2}+\cdots+n_{m_{q-1}}+1}^{n_{m_q}} \\ \delta_{n_{m_1}+n_{m_2}+\cdots+n_{m_{q-1}}+2}^{n_{m_q}} \\ \vdots \\ \delta_{n_{m_1}+n_{m_2}+\cdots+n_{m_q}}^{n_{m_q}} \end{pmatrix}$$

$$\Psi_s = \begin{pmatrix} y_{s_1} \\ \dot{y}_{s_1} \\ \vdots \\ y_{s_1}^{(n_{s_1}-1)} \\ y_{s_2} \\ \dot{y}_{s_2} \\ \vdots \\ y_{s_2}^{(n_{s_2}-1)} \\ \vdots \\ y_{s_r} \\ \dot{y}_{s_r} \\ \vdots \\ y_{s_q}^{(n_{s_r}-1)} \end{pmatrix} = \begin{pmatrix} \delta_1^{n_{s_1}} \\ \delta_2^{n_{s_1}} \\ \vdots \\ \delta_{n_{s_1}}^{n_{s_1}} \\ \delta_{n_{s_1}+1}^{n_{s_2}} \\ \delta_{n_{s_1}+2}^{n_{s_2}} \\ \vdots \\ \delta_{n_{s_1}+n_{s_2}}^{n_{s_2}} \\ \vdots \\ \delta_{n_{s_1}+n_{s_2}+\cdots+n_{s_{r-1}}+1}^{n_{s_r}} \\ \delta_{n_{s_1}+n_{s_2}+\cdots+n_{s_{r-1}}+2}^{n_{s_r}} \\ \vdots \\ \delta_{n_{s_1}+n_{s_2}+\cdots+n_{s_r}}^{n_{s_r}} \end{pmatrix}$$

By means of Ψ_m, the master family Σ_m can be represented as

$$\begin{aligned} \dot{\delta}_i^{n_{m_1}} &= \delta_{i+1}^{n_{m_1}}, \quad 1 \le i \le n_{m_1} - 1 \\ \dot{\delta}_{n_{m_1}}^{n_{m_1}} &= \mathscr{L}_{m_1}(\delta_1^{n_{m_1}}, \ldots, \delta_{n_{m_1}}^{n_{m_1}}) \\ \dot{\delta}_i^{n_{m_2}} &= \delta_{i+1}^{n_{m_2}}, \quad n_{m_1} + 1 \le i \le n_{m_1} + n_{m_2} - 1, \\ \dot{\delta}_{n_{m_1}+n_{m_2}}^{n_{m_2}} &= \mathscr{L}_{m_2}(\delta_{n_{m_1}+1}^{n_{m_2}}, \ldots, \delta_{n_{m_1}+n_{m_2}}^{n_{m_2}}) \\ &\vdots \\ \dot{\delta}_i^{n_{mq}} &= \delta_{i+1}^{n_{mq}}, \quad n_{m_1} + \cdots + n_{m_{q-1}} + 1 \le i \le n_{m_1} + \cdots + n_{m_q} - 1, \\ \dot{\delta}_{n_{m_1}+\cdots+n_{mq}}^{n_{mq}} &= \mathscr{L}_{m_q}(\delta_{n_{m_1}+\cdots+n_{m_{q-1}}+1}^{n_{mq}}, \ldots, \delta_{n_{m_1}+\cdots+n_{m_q}}^{n_{mq}}) \\ y_{m_j} &= \delta_l^{n_{m_j}}, \quad l = 1, n_{m_1} + 1, \ldots, n_{m_1} + n_{m_2} + \cdots + n_{m_{q-1}} + 1. \end{aligned} \tag{6.29}$$

In a compact form, the family (6.29) is

$$\begin{aligned} \dot{\delta}'_m &= \mathscr{A}'_m \delta'_m + \Phi'_m(\mathscr{L}_{m_1}, \ldots, \mathscr{L}_{m_q}) \\ \mathscr{Y}'_m &= \mathscr{C}'_m \delta'_m, \end{aligned} \tag{6.30}$$

where the master state vector family δ'_m is

$$\delta'_m = \left(\delta_1^{n_{m_1}}, \ldots, \delta_{n_{m_1}}^{n_{m_1}}, \ldots, \delta_{n_{m_1}+\cdots+n_{m_{q-1}}+1}^{n_{mq}}, \ldots, \delta_{n_{m_1}+\cdots+n_{mq}}^{n_{mq}}\right)^T$$

On the other hand, with Ψ_s we can represent Σ_s as

$$
\begin{aligned}
\dot{\delta}_i^{n_{s_1}} &= \delta_{i+1}^{n_{s_1}}, \quad 1 \le i \le n_{s_1} - 1 \\
\dot{\delta}_{n_{s_1}}^{n_{s_1}} &= \mathscr{L}_{s_1}(\delta_1^{n_{s_1}}, \ldots, \delta_{n_{s_1}}^{n_{s_1}}, u_{s_1}, \dot{u}_{s_1}, \ldots, u_{s_1}^{(\gamma_{s_1}-1)}) + u_{s_1}^{(\gamma_{s_1})} \\
\dot{\delta}_i^{n_{s_2}} &= \delta_{i+1}^{n_{s_2}}, \quad n_{s_1} + 1 \le i \le n_{s_1} + n_{s_2} - 1 \\
\dot{\delta}_{n_{s_1}+n_{s_2}}^{n_{s_2}} &= \mathscr{L}_{s_2}(\delta_{n_{s_1}+1}^{n_{s_2}}, \ldots, \delta_{n_{s_1}+n_{s_2}}^{n_{s_2}}, u_{s_2}, \dot{u}_{s_2}, \ldots, u_{s_2}^{(\gamma_{s_2}-1)}) + u_{s_2}^{(\gamma_{s_2})} \\
&\vdots \\
\dot{\delta}_i^{n_{s_r}} &= \delta_{i+1}^{n_{s_r}}, \quad n_{s_1} + \cdots + n_{s_{r-1}} + 1 \le i \le n_{s_1} + \cdots + n_{s_r} - 1, \\
\dot{\delta}_{n_{s_1}+\cdots+n_{s_r}}^{n_{s_r}} &= \mathscr{L}_{s_r}(\delta_{n_{s_1}+\cdots+n_{s_{r-1}}+1}^{n_{s_r}}, \ldots, \delta_{n_{s_1}+\cdots+n_{s_r}}^{n_{s_r}}, u_{s_r}, \dot{u}_{s_r}, \ldots, u_{s_r}^{(\gamma_{s_r}-1)}) \\
&\quad + u_{s_r}^{(\gamma_{s_r})} \\
y_{s_j} &= \delta_l^{n_{s_j}}, \quad l = 1, n_{s_1} + 1, \ldots, n_{s_1} + n_{s_2} + \cdots + n_{s_{r-1}} + 1
\end{aligned}
\tag{6.31}
$$

The family (6.31) in a compact form is

$$
\begin{aligned}
\dot{\delta}_s &= \mathscr{A}_s \delta_s + \Phi_s(\mathscr{L}_{s_1}, \ldots, \mathscr{L}_{s_r}) + \bar{\mathscr{U}}(u_{s_1}^{(\gamma_{s_1})}, u_{s_2}^{(\gamma_{s_2})}, \ldots, u_{s_r}^{(\gamma_{s_r})}) \\
\mathscr{Y}_s &= \mathscr{C}_s \delta_s
\end{aligned}
\tag{6.32}
$$

where the slave state vector family δ_s is

$$
\delta_s = \left(\delta_1^{n_{s_1}}, \ldots, \delta_{n_{s_1}}^{n_{s_1}}, \ldots, \delta_{n_{s_1}+\cdots+n_{s_{r-1}}+1}^{n_{s_r}}, \ldots, \delta_{n_{s_1}+\cdots+n_{s_r}}^{n_{s_r}} \right)^T
$$

Notice that each representation, (6.29) and (6.31), is a multi-output generalized observability canonical form (MGOCF). To proceed with our proof, we need to define the synchronization error as a difference between master and slave state vector families. Nevertheless, the slave family can be larger than the master family ($q < r$), such that there could be a dimension mismatch. Moreover, the interaction between the families is not clear. Therefore, we define the following transformation

$$
F = \begin{pmatrix} F_{11} & F_{12} & \ldots & F_{1r} \\ F_{21} & F_{22} & \ldots & F_{2r} \\ \vdots & \vdots & \ddots & \vdots \\ F_{q1} & F_{q2} & \ldots & F_{qr} \end{pmatrix} \in \mathbb{R}^{n_m \times n_s},
$$

where $F_{ij} = I$ if the ith master interacts with the jth slave. Otherwise, $F_{ij} = 0$. By applying this transformation to (6.30), we have

$$
\begin{aligned}
F\dot{\delta}'_m &= F\mathscr{A}'_m \delta'_m + F\Phi'_m(\mathscr{L}_{m_1}, \ldots, \mathscr{L}_{m_q}) \\
F\mathscr{Y}'_m &= F\mathscr{C}'_m \delta'_m,
\end{aligned}
$$

and after some algebraic manipulation, the compact form (6.30) is equivalent to

$$\begin{aligned} \dot{\delta}_m &= \mathscr{A}_m \delta_m + \Phi_m(\mathscr{L}_{m_1}, \ldots, \mathscr{L}_{m_q}) \\ \mathscr{Y}_m &= \mathscr{C}_m \delta_m, \end{aligned} \tag{6.33}$$

with $\mathscr{A}_m = \mathscr{A}_s = \mathscr{A}$, $\mathscr{C}_m = \mathscr{C}_s = \mathscr{C}$,

$$\delta_m = (\underbrace{\delta_{m_1}, \ldots, \delta_{m_1}}_{\varphi_{m_1} \text{ times}}, \underbrace{\delta_{m_2}, \ldots, \delta_{m_2}}_{\varphi_{m_2} \text{ times}}, \ldots, \underbrace{\delta_{m_q}, \ldots, \delta_{m_q}}_{\varphi_{m_q} \text{ times}})^T,$$

$$\Phi_m = (\underbrace{\phi_{m_1}(\mathscr{L}_{m_1}), \ldots, \phi_{m_1}(\mathscr{L}_{m_1})}_{\varphi_{m_1} \text{ times}}, \ldots, \underbrace{\phi_{m_q}(\mathscr{L}_{m_q}), \ldots, \phi_{m_q}(\mathscr{L}_{m_q})}_{\varphi_{m_q} \text{ times}})^T$$

where δ_{m_j}, $1 \leq j \leq q$ is the state vector of the jth master system. The indexes φ_j denote the number of systems from the slave family that interact with the jth master system. These indexes are such that $\varphi_{m_1} + \varphi_{m_2} + \cdots + \varphi_{m_q} = n_s$.

Then, let the synchronization error be

$$e_\delta = \delta_m - \delta_s \tag{6.34}$$

From (6.32) and (6.33) we obtain

$$\dot{e}_\delta = \mathscr{A} e_\delta + \Phi_m - \Phi_s - \bar{\mathscr{U}} \tag{6.35}$$

Such that we define $\bar{\mathscr{U}} = \Phi_m - \Phi_s + \mathscr{K} e_\delta$, where

$$\mathscr{K} = \begin{pmatrix} \mathrm{K}_1 & 0 & \ldots & 0 \\ 0 & \mathrm{K}_2 & \ldots & 0 \\ \vdots & \vdots & \ddots & \vdots \\ 0 & 0 & \ldots & \mathrm{K}_r \end{pmatrix}, \quad \mathrm{K}_j = \begin{pmatrix} 0 & 0 & \ldots & 0 \\ 0 & 0 & \ldots & 0 \\ \vdots & \vdots & & \vdots \\ 0 & 0 & \ldots & 0 \\ k_{1,j} & k_{2,j} & \ldots & k_{n_{s_j},j} \end{pmatrix},$$

$$1 \leq j \leq r.$$

Therefore, the family of dynamical controllers for the multi-synchronization is condensed as

$$
\begin{aligned}
\dot{e}_\delta &= \mathscr{A} e_\delta + \Phi_m - \Phi_s - \bar{\mathscr{U}} \\
\dot{\mathscr{U}} &= \bar{\mathscr{M}} \mathscr{U} + \bar{\mathscr{U}} \\
\bar{\mathscr{U}} &= \Phi_m - \Phi_s + \mathscr{K} e_\delta
\end{aligned} \tag{6.36}
$$

where

$$
\mathscr{U} = \begin{pmatrix} \mathscr{U}_{s_1} \\ \mathscr{U}_{s_2} \\ \vdots \\ \mathscr{U}_{s_r} \end{pmatrix}, \quad \mathscr{U}_{s_v} = \begin{pmatrix} u_{s_j} \\ \dot{u}_{s_j} \\ \vdots \\ u_{s_j}^{(\gamma_{s_j}-1)} \end{pmatrix}
$$

$$
\bar{\mathscr{M}} = \begin{pmatrix} M_1 & 0 & \dots & 0 \\ 0 & M_2 & \dots & 0 \\ \vdots & \vdots & \ddots & \vdots \\ 0 & 0 & \dots & M_r \end{pmatrix}, \quad M_j = \begin{pmatrix} 0\,1\,0 \dots 0 \\ 0\,0\,1 \dots 0 \\ \vdots\;\vdots\;\vdots\;\ddots\;\vdots \\ 0\,0\,0 \dots 1 \\ 0\,0\,0 \dots 0 \end{pmatrix}, \quad 1 \le j \le r.
$$

Then, in closed loop the dynamics of the synchronization error are

$$
\dot{e}_\delta = (\mathscr{A} - \mathscr{K}) e_\delta = \bar{\mathscr{A}} e_\delta \tag{6.37}
$$

where $\bar{\mathscr{A}}$ is

$$
\bar{\mathscr{A}} = \begin{pmatrix} \bar{\mathscr{A}}_1 & 0 & \dots & 0 \\ 0 & \bar{\mathscr{A}}_2 & \dots & 0 \\ \vdots & \vdots & \ddots & \vdots \\ 0 & 0 & \dots & \bar{\mathscr{A}}_r \end{pmatrix},
$$

$$
\bar{\mathscr{A}}_j = \begin{pmatrix} 0 & 1 & 0 & \dots & 0 \\ 0 & 0 & 1 & \dots & 0 \\ \vdots & \vdots & \vdots & \ddots & \vdots \\ 0 & 0 & 0 & \dots & 1 \\ -k_{1,j} & -k_{2,j} & -k_{3,j} & \dots & -k_{n_{s_j},j} \end{pmatrix}, \quad 1 \le j \le r.
$$

We can notice that the diagonal block matrix $\bar{\mathscr{A}}$ is a lineal operator and generates the strongly continuous semigroup $\exp(\bar{\mathscr{A}} t)$. Also, we know that its spectrum is

$$spec(\bar{\mathscr{A}}) = \bigcup_{j=1}^{r} spec(\bar{\mathscr{A}_j}) = \left\{\lambda_{1,j}, \lambda_{2,j}, \dots, \lambda_{n_{s_j},j}\right\}$$

Since we deal with families of integer order PDE-based systems, we can apply the stability analysis from the last section. We have that the spectral radius of the semigroup is

$$r(\exp(\bar{\mathscr{A}}t)) = \sup\left\{|\exp(\lambda_{1,j})|, |\exp(\lambda_{2,j})|, \dots, |\exp(\lambda_{n_{s_j},j})\right\}, \quad 1 \le j \le r.$$

Therefore, the solution of (6.37) will be exponentially stable if $Re(\lambda_{1,j}) < 0$, $Re(\lambda_{2,j}) < 0$,..., $Re(\lambda_{1,j}) < 0$ for $1 \le j \le r$. In such case, the families Σ_m and Σ_S are in a GMS state and the proof is completed. □

6.4.1 Synchronization of Multiple Flexible Body Aircraft

Let us consider multiple flexible body aircraft. Each one of them can be modeled as a hub-beam system (Fig. 6.14). Thus, the beam length is L and is embedded in the hub with radius R, whose rotation angle (attitude angle) is θ. We consider that the hub does not suffer any kind of deformation, such that $\omega(x, t)$ denotes the deformation on every point x of the beam. The signal u represents the actuator which makes the beam rotate. A hub-beam system is given by an integer order PDE-based system (see [27]). For our example, we consider two families of hub-beam systems given by

$$\begin{aligned}
\rho\ddot{P}_{m_\mu}(x,t) &= -EI\omega_{m_\mu xxxx}(x,t) \\
I_h\ddot{\theta}_{m_\mu}(t) &= EI\omega_{m_\mu xx}(0,t) \\
\omega_{m_\mu xxx}(L,t) &= 0 \\
\omega_{m_\mu xx}(L,t) &= 0 \\
\omega_{m_\mu x}(0,t) &= 0 \\
\omega_{m_\mu}(0,t) &= 0
\end{aligned} \tag{6.38}$$

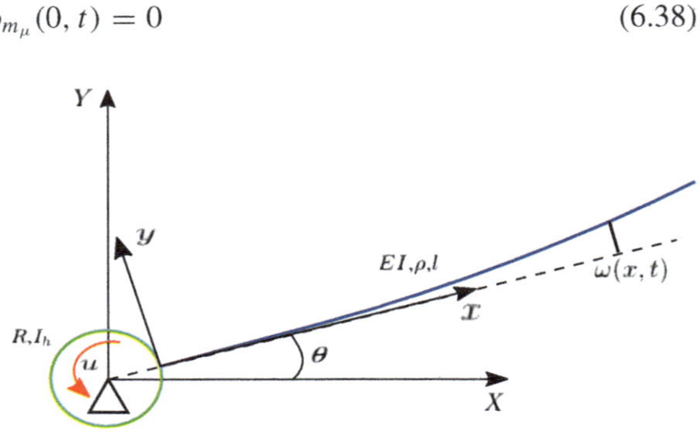

Fig. 6.14 Hub-beam system. Dynamics of each flexible-body aircraft

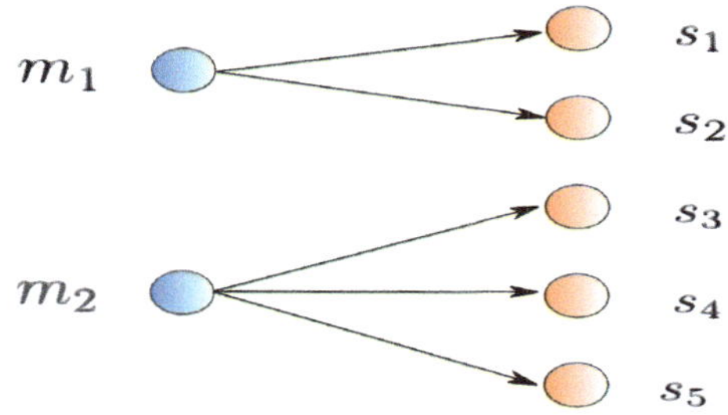

Fig. 6.15 Interaction of master and slave aircraft families

and

$$\begin{aligned}
\rho \ddot{P}_{s_\nu}(x,t) &= -EI\omega_{s_\nu xxxx}(x,t)\\
I_h\ddot{\theta}_{s_\nu}(t) &= EI\omega_{s_\nu xx}(0,t) + u_{s_\nu}\\
\omega_{s_\nu xxx}(L,t) &= 0\\
\omega_{s_\nu xx}(L,t) &= 0\\
\omega_{s_\nu x}(0,t) &= 0\\
\omega_{s_\nu}(0,t) &= 0
\end{aligned} \tag{6.39}$$

where $P(x,t)$ is the total position of x with respect to the inertial frame XY, this is, $P(x,t) = x\theta(t) + \omega(x,t)$. The constants ρ, EI and L are the linear density, stiffness and length of the beam. I_h is the rotary inertia of the hub. The family (6.38) is the master family and (6.39) is the salve family. These families interacted as is shown in Fig. 6.15. Our goal is that the attitude angle θ_{s_ν} of every slave is equal to its corresponding master θ_{m_μ}. We assume that θ is measurable for every aircraft.

Let the families of changes of variable be

$$y_{m_\mu} = \left\{\delta_1^{m_1}, \delta_3^{m_2}\right\} = \left\{\theta_{m_1}, \theta_{m_2}\right\}$$

and

$$\begin{aligned}
y_{s_\nu} &= \left\{\delta_1^{s_1}, \delta_3^{s_2}, \delta_5^{s_3}, \delta_7^{s_4}, \delta_9^{s_5}\right\}\\
&= \left\{\theta_{s_1} + u_1^{s_1}, \theta_{s_2} + u_1^{s_2}, \theta_{s_3} + u_1^{s_3}, \theta_{s_4} + u_1^{s_4}, \theta_{s_5} + u_1^{s_5}\right\},
\end{aligned}$$

where $u_1^{s_\nu} = u_{s_\nu}$. With these families, we can obtain the following families:

$$\begin{aligned}
\dot{\delta}_1^{m_1} &= \delta_2^{m_1}\\
\dot{\delta}_2^{m_1} &= \mathscr{L}_{m_1}(\delta_1^{m_1}, \delta_2^{m_1})\\
\dot{\delta}_3^{m_2} &= \delta_4^{m_2}\\
\dot{\delta}_4^{m_2} &= \mathscr{L}_{m_2}(\delta_3^{m_2}, \delta_4^{m_2})\\
y_{m_\mu} &= \delta_l^{m_\mu}, \quad l = 1, 3
\end{aligned} \tag{6.40}$$

where $\mathscr{L}_{m_\mu} = EI\omega_{m_\mu xx}(0,t)/I_h$. And

$$
\begin{aligned}
\dot{\delta}_1^{s1} &= \delta_2^{s1} \\
\dot{\delta}_2^{s1} &= \mathscr{L}_{s1}(\delta_1^{s1}, \delta_2^{s1}, u_1^{s1}, u_2^{s1}) + u_3^{s1} \\
\dot{\delta}_3^{s2} &= \delta_4^{s2} \\
\dot{\delta}_4^{s2} &= \mathscr{L}_{s2}(\delta_3^{s2}, \delta_4^{s2}, u_1^{s2}, u_2^{s2}) + u_3^{s2} \\
\dot{\delta}_5^{s3} &= \delta_6^{s3} \\
\dot{\delta}_6^{s3} &= \mathscr{L}_{s3}(\delta_5^{s3}, \delta_6^{s3}, u_1^{s3}, u_2^{s3}) + u_3^{s3} \\
\dot{\delta}_7^{s4} &= \delta_8^{s4} \\
\dot{\delta}_8^{s4} &= \mathscr{L}_{s4}(\delta_7^{s4}, \delta_8^{s4}, u_1^{s4}, u_2^{s4}) + u_3^{s4} \\
\dot{\delta}_9^{s5} &= \delta_{10}^{s5} \\
\dot{\delta}_{10}^{s5} &= \mathscr{L}_{s5}(\delta_9^{s5}, \delta_{10}^{s5}, u_1^{s5}, u_2^{s5}) + u_3^{s5} \\
y_{s_\nu} &= \delta_l^{s_\nu}, \quad l = 1, 3, 5, 7, 9.
\end{aligned}
\tag{6.41}
$$

where $\mathscr{L}_{s_\nu} = (EI\omega_{m_\mu xx}(0,t) + u_{s_\nu})/I_h$, $u_2^{s_\nu} = \dot{u}_1^{s_\nu}$ and $u_3^{s_\nu} = \dot{u}_2^{s_\nu}$. As we have more aircraft in the slave family than in the master family, let us define the transformation F which, from Fig. 6.14, is

$$
F = \begin{pmatrix} 1\,0\,1\,0\,0\,0\,0\,0\,0\,0 \\ 0\,1\,0\,1\,0\,0\,0\,0\,0\,0 \\ 0\,0\,0\,0\,1\,0\,1\,0\,1\,0 \\ 0\,0\,0\,0\,0\,1\,0\,1\,0\,1 \end{pmatrix}^T
$$

Therefore, the family (6.40) is equivalent to

$$
\begin{aligned}
\dot{\delta}_m &= \mathscr{A}\delta_m + \Phi_m(\mathscr{L}_{m_1}, \mathscr{L}_{m_2}) \\
\mathscr{Y}_m &= \mathscr{C}\delta_m,
\end{aligned}
\tag{6.42}
$$

with

$$
\delta_m = \left(\delta_1^{m_1}, \delta_2^{m_1}, \delta_1^{m_1}, \delta_2^{m_1}, \delta_3^{m_2}, \delta_4^{m_2}, \delta_3^{m_2}, \delta_4^{m_2}, \delta_3^{m_2}, \delta_4^{m_2}\right)^T,
$$

$$
\Phi_m(\mathscr{L}_{m_1}, \mathscr{L}_{m_2}) = \begin{pmatrix} \phi_{m_1}(\mathscr{L}_{m_1}) \\ \phi_{m_1}(\mathscr{L}_{m_1}) \\ \phi_{m_2}(\mathscr{L}_{m_2}) \\ \phi_{m_2}(\mathscr{L}_{m_2}) \\ \phi_{m_2}(\mathscr{L}_{m_2}) \end{pmatrix}, \quad \phi_{m_j}(\mathscr{L}_{m_j}) = \begin{pmatrix} 0 \\ \mathscr{L}_{m_j}(\delta_{2j-1}^{n_{m_j}}, \delta_{2j}^{n_{m_j}}) \end{pmatrix},
$$

$$
j = 1, 2.
$$

Remark 6.1. One can note that in the state vector family δ_m from (6.42), the state vector of the first master system repeats twice since the are two slave systems that interact with it. The same happens with the state vector of the second master system.

Meanwhile, the family (6.41) can be represented as

$$\begin{aligned}\dot{\delta}_s &= \mathscr{A}\delta_s + \Phi_s(\mathscr{L}_{s_1}, \mathscr{L}_{s_2}, \mathscr{L}_{s_3}, \mathscr{L}_{s_4}, \mathscr{L}_{s_5}) + \bar{\mathscr{U}},\\ \mathscr{Y}_s &= \mathscr{C}\delta_s,\end{aligned} \tag{6.43}$$

with

$$\delta^s = \left(\delta_1^{s_1}, \delta_2^{s_1}, \delta_3^{s_2}, \delta_4^{s_2}, \delta_5^{s_3}, \delta_6^{s_3}, \delta_7^{s_4}, \delta_8^{s_4}, \delta_9^{s_5}, \delta_{10}^{s_5}\right)^T$$

$$\Phi_s(\mathscr{L}_{s_1}, \ldots, \mathscr{L}_{s_5}) = \begin{pmatrix} \phi_{s_1}(\mathscr{L}_{s_1}) \\ \phi_{s_2}(\mathscr{L}_{s_2}) \\ \phi_{s_3}(\mathscr{L}_{s_3}) \\ \phi_{s_4}(\mathscr{L}_{s_4}) \\ \phi_{s_5}(\mathscr{L}_{s_5}) \end{pmatrix},$$

$$\phi_{s_j}(\mathscr{L}_{s_j}) = \begin{pmatrix} 0 \\ \mathscr{L}_{s_j}(\delta_{2j-1}^{n_{s_j}}, \delta_{2j}^{n_{s_j}}, u_1^{n_{s_j}}, u_2^{n_{s_j}}) \end{pmatrix},$$

$$\bar{\mathscr{U}}(u_3^{n_{s_1}}, \ldots, u_3^{n_{s_5}}) = \begin{pmatrix} \bar{\mathscr{U}}_{s_1}(u_3^{n_{s_1}}) \\ \bar{\mathscr{U}}_{s_2}(u_3^{n_{s_2}}) \\ \bar{\mathscr{U}}_{s_3}(u_3^{n_{s_3}}) \\ \bar{\mathscr{U}}_{s_4}(u_3^{n_{s_4}}) \\ \bar{\mathscr{U}}_{s_5}(u_3^{n_{s_5}}) \end{pmatrix}, \quad \bar{\mathscr{U}}_{s_j}(u_3^{n_{s_j}}) = \begin{pmatrix} 0 \\ u_3^{n_{s_j}} \end{pmatrix},$$

$$1 \leq j \leq 5.$$

For both families, matrices $\mathscr{A}$ and $\mathscr{C}$ are

$$\mathscr{A} = \begin{pmatrix} A_1 & 0 & 0 & 0 & 0 \\ 0 & A_2 & 0 & 0 & 0 \\ 0 & 0 & A_3 & 0 & 0 \\ 0 & 0 & 0 & A_4 & 0 \\ 0 & 0 & 0 & 0 & A_5 \end{pmatrix}, \quad A_j = \begin{pmatrix} 0 & 1 \\ 0 & 0 \end{pmatrix},$$

$$\mathscr{C}_s = \begin{pmatrix} C_1 & 0 & 0 & 0 & 0 \\ 0 & C_2 & 0 & 0 & 0 \\ 0 & 0 & C_3 & 0 & 0 \\ 0 & 0 & 0 & C_4 & 0 \\ 0 & 0 & 0 & 0 & C_5 \end{pmatrix}, \quad C_j = \begin{pmatrix} 1 & 0 \end{pmatrix}, \quad 1 \le j \le 5.$$

Therefore, from (6.42) and (6.43), the family of dynamical controllers is given by

$$\begin{aligned} \dot{e}_\delta &= \mathscr{A} e_\delta + \Phi_m - \Phi_s - \bar{\mathscr{U}} \\ \dot{\mathscr{U}} &= \bar{\mathscr{M}} \mathscr{U} + \bar{\mathscr{U}} \\ \bar{\mathscr{U}} &= \Phi_m - \Phi_s + \mathscr{K} e_\delta \end{aligned} \tag{6.44}$$

where

$$\mathscr{U} = \begin{pmatrix} \mathscr{U}_{s_1} \\ \mathscr{U}_{s_2} \\ \mathscr{U}_{s_3} \\ \mathscr{U}_{s_4} \\ \mathscr{U}_{s_5} \end{pmatrix}, \quad \mathscr{U}_{s_j} = \begin{pmatrix} u_1^{n_{s_j}} \\ u_2^{n_{s_j}} \end{pmatrix}$$

$$\bar{\mathscr{M}} = \begin{pmatrix} M_1 & 0 & 0 & 0 & 0 \\ 0 & M_2 & 0 & 0 & 0 \\ 0 & 0 & M_3 & 0 & 0 \\ 0 & 0 & 0 & M_4 & 0 \\ 0 & 0 & 0 & 0 & M_5 \end{pmatrix}, \quad M_j = \begin{pmatrix} 0 & 1 \\ 0 & 0 \end{pmatrix}, \quad 1 \le j \le 5.$$

Finally, in closed loop we have that $\dot{e}_\delta = \bar{\mathscr{A}} e_\delta$, with

$$\bar{\mathscr{A}} = \begin{pmatrix} \bar{\mathscr{A}}_1 & 0 & 0 & 0 & 0 \\ 0 & \bar{\mathscr{A}}_2 & 0 & 0 & 0 \\ 0 & 0 & \bar{\mathscr{A}}_3 & 0 & 0 \\ 0 & 0 & 0 & \bar{\mathscr{A}}_4 & 0 \\ 0 & 0 & 0 & 0 & \bar{\mathscr{A}}_5 \end{pmatrix}, \quad \bar{\mathscr{A}}_j = \begin{pmatrix} 0 & 1 \\ -k_{1,j} & -k_{2,j} \end{pmatrix}, \quad 1 \le j \le 5.$$

We know that the spectrum of $\bar{\mathscr{A}}$ is given by

$$\sigma(\bar{\mathscr{A}}) = \bigcup_{j=1}^{5} \sigma(\bar{\mathscr{A}}_j) = \left\{ \lambda_{1,j}, \lambda_{2,j} \right\},$$

Table 6.1 Initial attitude angle for masters and slaves

$\theta_{m_1} = 0\,\text{rad}$	$\theta_{m_2} = 1\,\text{rad}$
$\theta_{s_1}(0) = 0.3\,\text{rad}$	$\theta_{s_2}(0) = 0.2\,\text{rad}$
$\theta_{s_3}(0) = 0.2\,\text{rad}$	$\theta_{s_4}(0) = 0.4\,\text{rad}$
$\theta_{s_5}(0) = 0.6\,\text{rad}$	

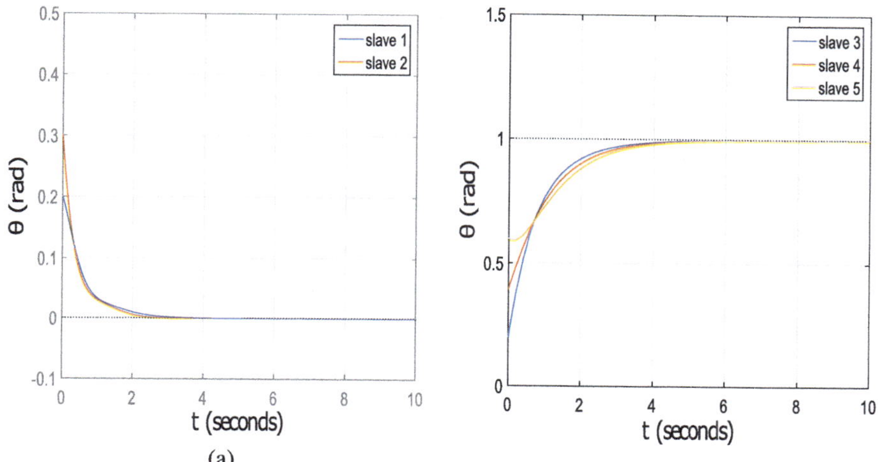

Fig. 6.16 GMS of master and slave families. Master aircraft 1 with slave aircraft 1 and 2 (**a**). Master aircraft 2 with slave aircraft 3, 4 and 5 (**b**)

where

$$\lambda_{1,j} = -\frac{1}{2}\left(k_{2,j} + \sqrt{-4k_{1,j} + k_{2,j}^2}\right),$$
$$\lambda_{2,j} = -\frac{1}{2}\left(k_{2,j} - \sqrt{-4k_{1,j} + k_{2,j}^2}\right), \quad 1 \leq j \leq 5.$$

Therefore, $k_{1,j} > 0$ and $k_{2,j} > 0$, since $Re(\lambda_{1,j}) < 0$ and $Re(\lambda_{2,j}) < 0$ for $1 \leq j \leq 5$.

For the numerical simulation, let us consider the initial conditions from Table 6.1, with fixed angles θ_{m_1} and θ_{m_2}. The gains are $k_{1,j} = k_{2,j} = 100$, $1 \leq j \leq 5$. The system parameters for each aircraft: $R = 1\,\text{m}$, $I_h = 0.5\,\text{kg m}^2$, $\rho = 0.02\,\text{kg/m}$, $L = 5\,\text{m}$ and $EI = 0.8\,\text{N m}^2$.

In Fig. 6.16 we observe how the attitude angle of each slave aircraft reaches its corresponding master's attitude angle. This is, the aircraft are in a MGS state.

References

1. Alinezhad, H.S., Yamchi, M.H., Esfanjani, R.M.: Robust synchronization of networked manipulators using distributed dynamic H_∞ controllers. ISA Trans. **83**, 239–247 (2018)
2. Beck, M.: A brief introduction to stability theory for linear PDEs. In: SIAM (2012)
3. Chen, B.-S., Chang, Y.-T.: Fuzzy state-space modeling and robust observer-based control design for nonlinear partial differential systems. IEEE Trans. Fuzzy Syst. **17**(5), 1025–1043 (2009)
4. Coron, J.-M., d'Andrea Novel, B., Bastin, G.: A strict Lyapunov function for boundary control of hyperbolic systems of conservation laws. IEEE Trans. Autom. Control **52**(1), 2–11 (2007)
5. De Wit, A., Lima, D., Dewel, G., Borckmans, P.: Spatiotemporal dynamics near a codimension-two point. Phys. Rev. E **54**(1), 261 (1996)
6. Demetriou, M.A.: Synchronization and consensus controllers for a class of parabolic distributed parameter systems. Syst. Control Lett. **62**(1), 70–76 (2013)
7. Demetriou, M.A., Fahroo, F.: Synchronization of a class of second order distributed parameter systems. IFAC Proc. Volumes **46**(26), 73–78 (2013)
8. Doelman, A., Kaper, T.J., Zegeling, P.A.: Pattern formation in the one-dimensional Gray-Scott model. Nonlinearity **10**(2), 523 (1997)
9. Feki, M.: An adaptive chaos synchronization scheme applied to secure communication. Chaos Solitons Fractals **18**(1), 141–148 (2003)
10. Fischer, I., Liu, Y., Davis, P.: Synchronization of chaotic semiconductor laser dynamics on subnanosecond time scales and its potential for chaos communication. Phys. Rev. A **62**(1), 011801 (2000)
11. Gray, P., Scott, S.: Autocatalytic reactions in the isothermal, continuous stirred tank reactor: isolas and other forms of multistability. Chem. Eng. Sci. **38**(1), 29–43 (1983)
12. Khadra, A., Liu, X., Shen, X.: Impulsive control and synchronization of spatiotemporal chaos. Chaos Solitons Fractals **26**(2), 615–636 (2005)
13. Mattheij, R.M., Rienstra, S.W., ten Thije Boonkkamp, J.H.: Partial differential equations: modeling, analysis, computation, vol. 10. In: Siam (2005)
14. Meng, Q., Shi, P.: Stochastic optimal control for backward stochastic partial differential systems. J. Math. Anal. Appl. **402**(2), 758–771 (2013)
15. Morris, K., Levine, W.: Control of systems governed by partial differential equations. In: The Control Theory Handbook (2010)
16. Oberkampf, W.L., DeLand, S.M., Rutherford, B.M., Diegert, K.V., Alvin, K.F.: Error and uncertainty in modeling and simulation. Reliab. Eng. Syst. Saf. **75**(3), 333–357 (2002)
17. Smyshlyaev, A., Krstic, M.: Closed-form boundary state feedbacks for a class of 1-d partial integro-differential equations. IEEE Trans. Autom. Control **49**(12), 2185–2202 (2004)
18. Solis, M.A., Olivares, M., Allende, H.: Stabilizing dynamic state feedback controller synthesis: a reinforcement learning approach. Stud. Inf. Control **25**(2), 245–254 (2016)
19. Tai, W., Teng, Q., Zhou, Y., Zhou, J., Wang, Z.: Chaos synchronization of stochastic reaction-diffusion time-delay neural networks via non-fragile output-feedback control. Appl. Math. Comput. **354**, 115–127 (2019)
20. Wang, J., Wu, K.-N., Pan, P.-L.: Finite-time synchronization of coupled stochastic partial differential systems. In: 2015 34th Chinese Control Conference (CCC), pp. 1705–1709. IEEE (2015)
21. Wang, K., Steyn-Ross, M.L., Steyn-Ross, D.A., Wilson, M.T., Sleigh, J.W., Shiraishi, Y.: Simulations of pattern dynamics for reaction-diffusion systems via simulink. BMC Syst. Biol. **8**(1), 45 (2014)
22. Wu, K., Chen, B.-S.: Synchronization of partial differential systems via diffusion coupling. IEEE Trans. Circuits Syst. I Regul. Pap. **59**(11), 2655–2668 (2012)
23. Wu, K.-N., Li, C.-X., Chen, B.-S., Yao, Y.: Robust H_∞ synchronization of coupled partial differential systems with spatial coupling delay. IEEE Trans. Circuits Syst. II Express Briefs **60**(7), 451–455 (2013)
24. Wu, K.-N., Tian, T., Wang, L.: Synchronization for a class of coupled linear partial differential systems via boundary control. J. Franklin Inst. **353**(16), 4062–4073 (2016)

25. Wu, K.-N., Tian, T., Wang, L., Wang, W.-W.: Asymptotical synchronization for a class of coupled time-delay partial differential systems via boundary control. Neurocomputing **197**, 113–118 (2016)
26. Yang, C., Cheng, L., Sun, K., Chen, X., Li, T., Wang, Y., Chen, X., Zhang, A., Qiu, J.: Asymptotical synchronization of a class of driving-response pde networks with time delay and spatially variable coefficients. In: 2015 11th International Conference on Natural Computation (ICNC), pp. 530–534. IEEE (2015)
27. Zou, A.-M., de Ruiter, A.H., Kumar, K.D.: Distributed attitude synchronization control for a group of flexible spacecraft using only attitude measurements. Inf. Sci. **343**, 66–78 (2016)

Chapter 7
Synchronization and Fractional-Order Systems

In this chapter, in a master–slave configuration of strictly different commensurate *fractional-order* Liouvillian systems, the generalized synchronization problem of multiple decoupled families of Liouvillian systems is addressed. Once again, the key ingredient is to find canonical forms for the original systems, from a family of fractional differential primitive elements and taking into account the Liouvillian feature. A set of *fractional-order dynamical controllers* is designed to solve the generalized multi-synchronization problem. Moreover, it is shown that adding diffusive coupling terms in the dynamical controllers solves the synchronization problem with a complex interaction between slave systems, with any type of interplay. In addition, as an extension of PDE systems of integer order, the problem of generalized synchronization for PDE systems of fractional order is also addressed in this chapter.

7.1 Fractional Systems and the Synchronization Problem

Fractional calculus is the branch of mathematics that studies integrals and derivatives of arbitrary order, such as $\partial^{1/2} f(x,t)/\partial t^{1/2}$. It has existed almost as long as classical calculus. The first mention dates from 1695 when Leibnitz questioned the meaning of the semi-derivative in a letter written to l'Hôpital [14]. It is said that fractional calculus is a generalization of classical calculus since many classical definitions can be recovered when the fractional derivative order is replaced by an integer number. Although there have been some attempts, fractional calculus still lacks physical and geometrical interpretations accepted by the scientific community (see [16, 27]). However, it has been applied with success in multiple disciplines, in particular in those where memory effects and hereditary properties are relevant. For example, in a visco-elastic media, the stress at a particular time depends on the strains not only at that time but also on the history of the process. Nowadays, fractional calculus

R. Martínez-Guerra and J. P. Flores-Flores, *An Approach to Multi-agent Systems as a Generalized Multi-synchronization Problem*, Understanding Complex Systems,
https://doi.org/10.1007/978-3-031-22669-4_7

is widely used to model the rheological behavior of materials such as polymers or brain tissue [13, 17, 35], as well as anomalous diffusion and advection processes [10, 46].

Recently, the synchronization of *fractional-order chaotic systems* has received much attention as the integer-order case. The former is a broad topic tackled from different techniques: A one-way coupling and a projective synchronization scheme for the unified fractional-order system are addressed in [37, 45], respectively. Dynamical analysis for the one-way coupling scheme of fractional-order Liu systems is obtained in [41]. Synchronization of hyperchaotic Lorenz system is tackled from an active control technique in [40]. In [22] a linear active control technique is used for synchronization in driver and response configuration, where the proposed methodology is applied to synchronize identical systems with commensurate and incommensurate fractional order. An active sliding mode control is given in [28, 34] , and a modified version of [34] is given in terms of the projective synchronization problem in [38] . An adaptive projective synchronization method with parametric uncertainty of fractional Lorenz systems with a reduced number of active control signals is given in [1], and it is proven that synchronization errors can only be bounded. In [3] a synchronization method with optimal active control and the fractional cost function is proposed. Particularly in [8, 19], both consider the generalized synchronization problem for fractional-order chaotic systems.

The synchronization of a pair of systems has been extended to the study of more complex problems involving multiple systems, of course, motivated by problems in the integer-order case where synchronization is observed such as rendezvous, formation control, flocking and schooling, attitude alignment, sensor networks, distributed computing, consensus and complex networks in general [9, 11, 15, 18, 24, 29, 32, 42]. In the fractional-order case, we can mention as related work from a control theory perspective: a multi-synchronization scheme for identical systems in a ring connection with unidirectional and bidirectional coupling [7]. In [36] a pinning synchronization problem is presented for a network of systems with Lipschitz-type nonlinearities and unidirectional configuration. In terms of generalized synchronization, a modification of active control is given in [44] where it is considered the case of interaction between multiple slave systems.

In recent years, active control has been a popular technique used in chaos synchronization. These controllers use an static control signal for each first-order fractional differential equation [39], and this suggests the search for appropriate signals to obtain stable error dynamics. The last requires a great amount of effort; therefore, this chapter presents a reduced number of fractional dynamical control laws able to stabilize the origin of error dynamics. This methodology synchronizes a class of multiple chaotic decoupled nonlinear fractional-order systems, where systems are not necessarily identical and is sufficient to construct a family of fractional differential primitive elements that generates a family of transformations. Thus, systems are carried out to a *multi-output fractional generalized observability canonical form* (MFGOCF).

Finally, from this coordinate transformation, the synchronization of multiple decoupled families of chaotic systems in a master–slave configuration is possible. This type of synchronization is introduced as *fractional generalized multi-synchronization* (FGMS). Moreover, it is also considered the case of a complex interaction between slave systems as a natural extension of the former result, as it will be shown that any type of interplay between slave systems can be considered and still obtain synchronization error convergence to the origin.

7.1.1 Fractional Calculus Preliminaries

A large amount of definitions about fractional-order differentiation and integration operators are available in literature [23, 26]. Here, the *Caputo fractional derivative* will be used, since for a fractional differential equation the initial conditions can be given in terms of differential quantities with a physical meaning. Moreover, unlike most of the fractional differentiation operators, the Caputo derivative of a constant is zero, as in integer-order case.

Definition 7.1. The Caputo fractional derivative of order $\alpha \in \mathbb{R}$ of a function $x(t)$ is defined as

$$x^{(\alpha)}(t) = {}_{t_0}D_t^{\alpha} x(t) = \frac{1}{\Gamma(n-\alpha)} \int_{t_0}^{t} x^{(n)}(\tau)(t-\tau)^{n-\alpha-1} d\tau \tag{7.1}$$

where $n \in \mathbb{N}$, $n-1 < \alpha < n$ and $x^{(n)}(\tau)$ is the n-th derivative of $x(\tau)$ in the usual sense, $n \in \mathbb{N}$ and Γ is the gamma function, defined as

$$\Gamma(\alpha) = \int_{0}^{\infty} t^{\alpha-1} e^{-t} dt.$$

An extensive review of the gamma function and its properties can be found in [2].

There exists a fractional generalization of the *Fundamental Theorem of Calculus* for finite intervals. This is the case when the integral and derivative operators are taken as the Riemann-Liouville integral and the Caputo derivative, respectively [33].

Definition 7.2. The Riemann-Liouville fractional integral of order $\alpha \in \mathbb{R}$ of a function $x(t)$ is defined as

$$x^{(-\alpha)}(t) = {}_{t_0}I_t^{\alpha} x(t) = \frac{1}{\Gamma(\alpha)} \int_{t_0}^{t} x(\tau)(t-\tau)^{\alpha-1} d\tau \tag{7.2}$$

where $0 \leq \alpha \leq 1$ and Γ is the gamma function.

To simplify the notation, time dependence of x will be omitted. Moreover, in what follows $t_0 = 0$. Let us define the sequential operator

$$\mathscr{D}^{r\alpha}x(t) = \underbrace{{}_{t_0}D_t^{\alpha}\,{}_{t_0}D_t^{\alpha}\cdots{}_{t_0}D_t^{\alpha}\,{}_{t_0}D_t^{\alpha}}_{r-times}x(t) \tag{7.3}$$

this is the Caputo fractional derivative of order α applied $r \in \mathbb{N}$ times sequentially. One can note that $\mathscr{D}^0 x(t) = x(t)$, and with $r = 1$ then $\mathscr{D}^{\alpha}x(t) = x^{(\alpha)}$. From now the operator for derivation and integration will be defined as in (7.1) and (7.2), respectively. Consider the following dynamical system

$$\begin{aligned}\mathscr{D}^{\alpha}x &= F(x,u)\\ y &= h(x,u)\end{aligned} \tag{7.4}$$

where $x = (x_1, x_2, \ldots, x_n) \in \mathbb{R}^n$, $u = (u_1, u_2, \ldots, u_{\gamma}) \in \mathbb{R}^{\gamma}$, $F : \mathbb{R}^n \times \mathbb{R}^{\gamma} \to \mathbb{R}^n$, $h : \mathbb{R}^n \to \mathbb{R}$.

Definition 7.3. A state variable $x_i \in \mathbb{R}$ of system (7.4) satisfies the Fractional Algebraic Observability (FAO) if x_i is a function of the first $r_1, r_2 \in \mathbb{N}$ sequential fractional derivatives of the available output y and the input u, respectively, i.e.,

$$x_i = \phi_i\left(y, y^{(\alpha)}, \mathscr{D}^{2\alpha}y, \ldots, \mathscr{D}^{r_1\alpha}y, u, u^{(\alpha)}, \mathscr{D}^{2\alpha}u, \ldots, \mathscr{D}^{r_2\alpha}u\right)$$

where $\phi_i : \mathbb{R}^{(r_1+1)p} \times \mathbb{R}^{(r_2+1)m} \to \mathbb{R}$.

There exist some systems that do not necessarily satisfy FAO. Then, the following definition is given.

Definition 7.4. Let $\bar{n}$ states of system (7.4) satisfy FAO property for $\bar{n} < n$, then we will say that system (7.4) is Fractional Liouvillian if the $n - \bar{n}$ states can be obtained by adjunction of fractional-order integrals of the $\bar{n}$ states.

As a result of the above definition, we can modify Definition 7.3 as follows

Definition 7.5. A state variable $x_i \in \mathbb{R}$ satisfies Fractional Liouvillian Algebraic Observability (FLAO) if x_i is a function of the first $r_1, r_2 \in \mathbb{N}$ sequential fractional derivatives of the available output $\bar{y} = I^{\alpha}y$ and the input u, respectively, i.e.,

$$x_i = \phi_i\left(I^{\alpha}y, y, \mathscr{D}^{\alpha}y, \ldots, \mathscr{D}^{(r_1-1)\alpha}y, \ldots, u, u^{(\alpha)}, \mathscr{D}^{2\alpha}u, \ldots, \mathscr{D}^{r_2\alpha}u\right)$$

where $\phi_i : \mathbb{R}^{(r1+1)p} \times \mathbb{R}^{(r2+1)m} \to \mathbb{R}$.

From the above, Definition 2.5 can be extended as

Definition 7.6. A family of systems is Picard-Vessiot (PV) if and only if the vector space generated by the fractional derivatives of the family

$$\left\{\mathscr{D}^{n_j\alpha}\bar{y}_j, n_j \geq 0, 1 \leq j \leq q, 0 \leq \alpha \leq 1\right\}$$

has finite dimension, where $\bar{y}_j$ is the j-th output (fractional differential primitive element).

Theorem 7.1. ([20]) *Let* $\alpha < 2$ *and* $\bar{A} \in \mathbb{C}^{n\times n}$. *The autonomous system*

$$x^{(\alpha)} = \bar{A}x, \quad x(0) = x_0$$

is asymptotically stable if and only if $|arg(\lambda_i(\bar{A}))| > \alpha\pi/2$, *where* $\lambda_i(\bar{A})$ *is the* i*-th eigenvalue of the matrix* $\bar{A}$.

Remark 7.1. As a particular case from above theorem, for $0 < \alpha < 1$, all Hurwitz matrix satisfy the condition

$$|arg(\lambda_i(\bar{A}))| > \frac{\pi}{2} > \frac{\alpha\pi}{2}$$

7.2 Generalized Synchronization for Families of Fractional Systems

The synchronization of families of decoupled nonlinear fractional-order systems can also be reduced to multiple fractional generalized synchronization problems where is sufficient to know the output of each system to generate a family of transformations which give us the possibility to synchronize multiple chaotic systems, and these transformations are obtained from a family of outputs given by $\bar{y}_j = I^\alpha y_j$ with $1 \leq j \leq p$ (p outputs). Let $n_j \geq 0$ be the minimum integers such that $\mathscr{D}^{n_j\alpha}\bar{y}_j$ are analytically dependent on $(\bar{y}_j, \bar{y}_j^{(\alpha)}, \ldots, \mathscr{D}^{[n_j-1]\alpha}\bar{y}_j)$, where $\bar{y}_j = I^\alpha y_j$:

$$\bar{H}_j(\bar{y}_j, \bar{y}_j^{(\alpha)}, \ldots, \mathscr{D}^{[n_j-1]\alpha}\bar{y}_j, \mathscr{D}^{n_j\alpha}\bar{y}_j, u_j, u_j^{(\alpha)}, \ldots, \mathscr{D}^{\gamma_j\alpha}u_j) = 0 \tag{7.5}$$

The system (7.5) can be solved locally as:

$$\mathscr{D}^{n_j\alpha}\bar{y}_j = -\mathscr{L}_j(\bar{y}_j, \bar{y}_j^{(\alpha)}, \ldots, \mathscr{D}^{[n_j-1]\alpha}\bar{y}_j, u_j, u_j^{(\alpha)}, \ldots, \mathscr{D}^{[\gamma_j-1]\alpha}u_j) + \mathscr{D}^{\gamma_j\alpha}u_j$$

Let $\xi_i^{n_j} = \mathscr{D}^{[i-l]\alpha}\bar{y}_j, l = 1, n_1+1, n_1+n_2+1, \ldots, n_1+n_2+\cdots+n_{p-1}+1;$ $1 \leq i \leq \sum_{1\leq j\leq p} n_j = n$ where index j gives the j-th system and the n_j's are the so-called indices of algebraic observability where each index coincides with the system's dimension. Then it is possible to achieve a local representation for a set of p decoupled systems, and this representation can be seen as a multi-output fractional generalized observability canonical form (MFGOCF):

$$
\begin{aligned}
\mathscr{D}^{\alpha}\xi_{1}^{n_1} &= \xi_{2}^{n_1}\\
\mathscr{D}^{\alpha}\xi_{2}^{n_1} &= \xi_{3}^{n_1}\\
&\vdots\\
\mathscr{D}^{\alpha}\xi_{n_1-1}^{n_1} &= \xi_{n_1}^{n_1}\\
\mathscr{D}^{\alpha}\xi_{n_1}^{n_1} &= -\mathscr{L}_1(\xi_{1}^{n_1},\xi_{2}^{n_1},\ldots,\xi_{n_1}^{n_1},u_1,u_1^{(\alpha)},\ldots,\mathscr{D}^{[\gamma_1-1]\alpha}u_1)+\mathscr{D}^{\gamma_1\alpha}u_1\\
\mathscr{D}^{\alpha}\xi_{n_1+1}^{n_2} &= \xi_{n_1+2}^{n_2}\\
\mathscr{D}^{\alpha}\xi_{n_1+2}^{n_2} &= \xi_{n_1+3}^{n_2}\\
&\vdots\\
\mathscr{D}^{\alpha}\xi_{n_1+n_2-1}^{n_2} &= \xi_{n_1+n_2}^{n_2}\\
\mathscr{D}^{\alpha}\xi_{n_1+n_2}^{n_2} &= -\mathscr{L}_2(\xi_{n_1+1}^{n_2},\xi_{n_1+2}^{n_2},\ldots,\xi_{n_1+n_2}^{n_2},u_2,u_2^{\alpha},\ldots,\mathscr{D}^{[\gamma_2-1]\alpha}u_2)+\mathscr{D}^{\gamma_2\alpha}u_2\\
&\vdots\\
\mathscr{D}^{\alpha}\xi_{n_1+n_2+\cdots+n_{p-1}+1} &= \xi_{n_1+n_2+\cdots+n_{p-1}+2}^{n_p}\\
\mathscr{D}^{\alpha}\xi_{n_1+n_2+\cdots+n_{p-1}+2} &= \xi_{n_1+n_2+\cdots+n_{p-1}+3}^{n_p}\\
&\vdots\\
\mathscr{D}^{\alpha}\xi_{n_1+n_2+\cdots+n_{p-1}+n_p-1}^{n_p} &= \xi_{n_1+n_2+\cdots+n_{p-1}+n_p}^{n_p}\\
\mathscr{D}^{\alpha}\xi_{n_1+n_2+\cdots+n_{p-1}+n_p}^{n_p} &= -\mathscr{L}_p(\xi_{n_1+n_2+\cdots+n_{p-1}+1}^{n_p},\xi_{n_1+n_2+\cdots+n_{p-1}+2}^{n_p},\ldots,\\
&\quad\ \xi_{n_1+n_2+\cdots+n_{p-1}+n_p}^{n_p},u_p,u_p^{\alpha},\ldots,\mathscr{D}^{[\gamma_p-1]\alpha}u_p)+\mathscr{D}^{\gamma_p\alpha}u_p\\
y_j &= \xi_{l}^{n_j}
\end{aligned}
\tag{7.6}
$$

In a compact form the new system (7.6) can be represented as:

$$
\begin{aligned}
\mathscr{D}^{\alpha}\xi &= \mathscr{A}\xi-\Phi(\mathscr{L}_1,\ldots,\mathscr{L}_p)+\bar{\mathscr{U}}(\mathscr{D}^{\gamma_1\alpha}u_1,\ldots,\mathscr{D}^{\gamma_p\alpha}u_p)\\
\mathscr{Y} &= \mathscr{C}\xi
\end{aligned}
\tag{7.7}
$$

where ξ, Φ, $\bar{\mathscr{U}}\in\mathbb{R}^n$, $\mathscr{A}\in\mathbb{R}^{n\times n}$, $\mathscr{Y}\in\mathbb{R}^p$ and the matrices of (7.7) are defined as follows:

$$
\mathscr{A}=\begin{bmatrix} A_1 & & 0\\ & \ddots & \\ 0 & & A_p\end{bmatrix};
$$

$$A_j = \begin{bmatrix} 0\,1\,0 & 0 & \ldots & 0 \\ 0\,0\,1 & 0 & \ldots & 0 \\ \vdots\,\vdots\,\vdots & \ddots & \ldots & 0 \\ 0\,0\,0 & 0 & 1 & 0 \\ 0\,0\,0 & 0 & 0 & 1 \\ 0\,0\,0 & 0 & 0 & 0 \end{bmatrix}, \quad 1 \leq j \leq p;$$

$$\Phi(\mathscr{L}_1, \ldots, \mathscr{L}_p) = \begin{bmatrix} \phi_1(\mathscr{L}_1) \\ \phi_2(\mathscr{L}_1) \\ \vdots \\ \phi_p(\mathscr{L}_p) \end{bmatrix};$$

$$\phi_j(\mathscr{L}_j) = \begin{bmatrix} 0 \\ 0 \\ \vdots \\ 0 \\ -\mathscr{L}_j(\xi^{n_j}_{n_1+n_2+\cdots+n_{j-1}+1}, \ldots, \xi^{n_p}_{n_1+n_2+\cdots+n_j}, u_j, u_j^{(\alpha)}, \ldots, \mathscr{D}^{[\gamma_j-1]\alpha} u_j) \end{bmatrix};$$

$$\bar{\mathscr{U}}(\mathscr{D}^{\gamma_1\alpha} u_1, \ldots, \mathscr{D}^{\gamma_p\alpha} u_p) = \begin{bmatrix} u_1(\mathscr{D}^{\gamma_1\alpha} u_1) \\ u_2(\mathscr{D}^{\gamma_2\alpha} u_2) \\ \vdots \\ u_p(\mathscr{D}^{\gamma_p\alpha} u_p) \end{bmatrix};$$

$$u_j(\mathscr{D}^{\gamma_j\alpha} u_j) = \begin{bmatrix} 0 \\ 0 \\ \vdots \\ 0 \\ \mathscr{D}^{\gamma_j\alpha} u_j \end{bmatrix};$$

$$\mathscr{C} = \begin{bmatrix} C_1 & & 0 \\ & \ddots & \\ 0 & & C_p \end{bmatrix};$$

$$C_j = \begin{bmatrix} 1 \; 0 \ldots 0 \end{bmatrix}.$$

Now, consider the following family of chaotic nonlinear systems

$$\begin{aligned} x_j^{(\alpha)} &= F_j(x_j, u_j) \\ y_j &= C_j x_j + D_j u_j \end{aligned} \tag{7.8}$$

where $1 \leq j \leq p$ denotes the j-th system, $x_j \in \mathbb{R}^{n_j}$ is the state vector, $F_j(\cdot)$ is a nonlinear vector function, u_j is the input, y_j is the output and C_j, D_j are matrices of appropriate size.

Lemma 7.1. *Consider the family of nonlinear systems* (7.8). *If the output is chosen as*

$$y_j = \sum_{i=n-n_j+1}^{n} \gamma_i x_i + \sum_{k}^{m} \beta_k u_k,$$

where γ_i, β_k are differential quantities of u and their time finite derivatives, such that the first component of the coordinate transformation is given by $\bar{y}_j = I^\alpha y_j$; then the nonlinear system (7.8) *is transformable to a MFGOCF if and only if is a family of PV systems.*

Proof. Let the set $\left\{\zeta_j, \zeta_j^{(\alpha)}, \ldots, \mathscr{D}^{[n_j-1]\alpha}\zeta_j\right\}$, $1 \leq j \leq p$ with $\zeta_j = I^{(\alpha)} y_j = \bar{y}_j$, $\mathscr{D}^{[i-l]\alpha}\zeta_j = \mathscr{D}^{[i-l-1]\alpha} y_j$, $1 \leq i \leq \sum_{j=1}^{p} n_j$, where $n_j \geq 0$ is the minimum integer such that $\mathscr{D}^{[n_j-1]\alpha} y_j$ is dependent on $I^{(\alpha)} y_j, y, y_j^{(\alpha)}, \ldots, \mathscr{D}^{[n_j-l-1]\alpha} y_j, u_j, \ldots$

Then, by redefining $\xi_i = \zeta_j = I^{(\alpha)} y_j, \xi_i^{n_j} = \mathscr{D}^{[i-l]\alpha}\zeta_j = \mathscr{D}^{[i-l-1]\alpha} y_j$, $1 \leq i \leq \sum_{j=1}^{p} n_j$ that yields to:

$$\begin{aligned}
\mathscr{D}^\alpha \xi_i^{n1} &= \xi_{i+1}^{n1}, \quad 1 \leq i \leq n_1 - 1 \\
\mathscr{D}^\alpha \xi_{n_1}^{n1} &= -\mathscr{L}_1(\xi_1^{n1}, \ldots, \xi_{n_1}^{n1}, u_1, u_1^{(\alpha)}, \ldots, \mathscr{D}^{[\gamma_1-1]\alpha} u_1) + \mathscr{D}^{\gamma_1\alpha} u_1 \\
\mathscr{D}^\alpha \xi_i^{n2} &= \xi_{i+1}^{n2}, \quad n_1 + 1 \leq i \leq n_1 + n_2 - 1 \\
\mathscr{D}^\alpha \xi_{n_1+n_2}^{n2} &= -\mathscr{L}_2(\xi_{n_1+1}^{n2}, \ldots, \xi_{n_1+n_2}^{n2}, u_2, u_2^{(\alpha)}, \ldots, \mathscr{D}^{[\gamma_2-1]\alpha} u_2) + \mathscr{D}^{\gamma_2\alpha} u_2 \\
&\vdots \\
\mathscr{D}^\alpha \xi_i^{np} &= \xi_{i+1}^{np}, \quad n_1 + \cdots + n_{p-1} + 1 \leq i \leq n_1 + \cdots + n_p - 1 \\
\mathscr{D}^\alpha \xi_{n_1+\cdots+n_p}^{np} &= -\mathscr{L}_p(\xi_{n_1+\cdots+n_{p-1}+1}^{np}, \ldots, \xi_{n_1+\cdots+n_p}^{np}, u_p, u_p^{(\alpha)}, \ldots, \mathscr{D}^{[\gamma_p-1]\alpha} u_p) + \mathscr{D}^{\gamma_p\alpha} u_p \\
\xi_l^{n_j} &= I^\alpha y_j, \quad 1 \leq j \leq p, l = 1 + \sum_{\hat{j}=1}^{j-1} n_{\hat{j}}.
\end{aligned}$$

and the proof is completed. □

Now is discussed the problem of generalized synchronization for a class of fractional-order systems, the so-called Fractional Liouvillian Systems. Within this

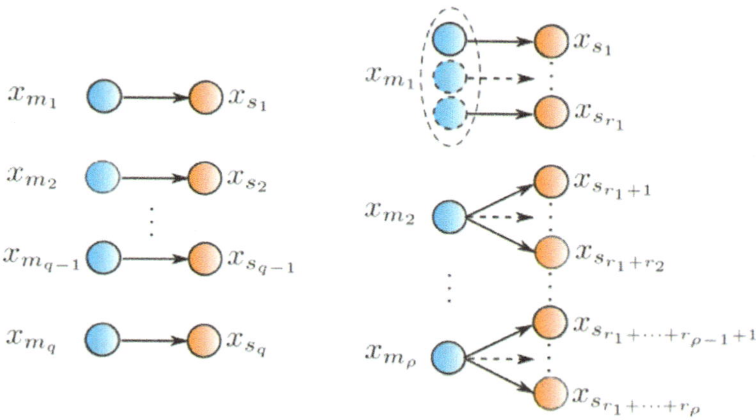

Fig. 7.1 Generalized Multi-synchronization configuration: an equal number of slaves and masters (left), more slaves than masters (right)

class is possible to find some fractional-order chaotic systems. In this case, let us consider a master–slave configuration. Define a family of master systems as:

$$\begin{aligned} \mathscr{D}^{(\alpha)} x_{m_\mu} &= F_{m_\mu}(x_{m_\mu}, u_{m_\mu}) \\ y_{m_\mu} &= h_{m_\mu}(x_{m_\mu}, u_{m_\mu}) \end{aligned} \tag{7.9}$$

and the family of slave systems:

$$\begin{aligned} \mathscr{D}^{(\alpha)} x_{s_\nu} &= F_{s_\nu}(x_{s_\nu}, u_{s_\nu}) \\ y_{s_\nu} &= h_{s_\nu}(x_{s_\nu}, u_{s_\nu}) \end{aligned} \tag{7.10}$$

where $x_{s_\nu} = (x_{1,s_\nu}, \ldots, x_{n_{s_\nu},s_\nu}) \in \mathbb{R}^{n_{s_\nu}}$, $x_{m_\mu} = (x_{1,m_\mu}, \ldots, x_{n_{m_\mu},m_\mu}) \in \mathbb{R}^{n_{m_\mu}}$, $h_{m_\mu} : \mathbb{R}^{n_{m_\mu}} \to \mathbb{R}$, $u_{m_\mu} = (u_{1,m_\mu}, \ldots, u_{\gamma_{m_\mu},m_\mu}) \in \mathbb{R}^{\gamma_{m_\mu}}$, $u_{s_\nu} = (u_{1,s_\nu}, \ldots, u_{\gamma_{s_\nu},s_\nu}) \in \mathbb{R}^{\gamma_{s_\nu}}$, $1 \le \nu \le p-1$, $1 \le \mu \le p-\nu$; these conditions tell us that one or more slave systems can be associated with one master but a slave is associated only with one master. One slave is associated with one master when the number of slaves is equal to the number of masters; on the other hand, it is possible to consider a case where the number of slaves is greater than the number of masters, and this means that a master system interacts with more than one slave system. This configuration is depicted in Fig. 7.1, and the circles or nodes represent the dynamic systems involved. Encircled nodes in Fig. 7.1 represent the same master system, dashed circles represent virtual master systems, and these possess the same dynamics and initial conditions as the original master system (solid circles) associated with the corresponding slave system; hence it can be represented as a single master node.

Definition 7.7. Let the vectors $X_m = (x_{m_1}, \ldots, x_{n_m}) \in \mathbb{R}^{n_m}$ and $X_s = (x_{s_1}, \ldots, x_{n_s}) \in \mathbb{R}^{n_s}$ be master and slave state vectors families, respectively; then the family of slave systems is in a state of fractional generalized multi-synchronization (FGMS) with their family of master systems if there exists a family of fractional outputs that generates a transformation $H_{ms} : \mathbb{R}^{n_s} \rightarrow \mathbb{R}^{n_m}$ with $H_{ms} = \phi_m^{-1} \circ \phi_s$ and there exist an algebraic manifold $M = \{(X_s, X_m) | X_m = H_{ms}(X_s)\}$ and a compact set $B \subset \mathbb{R}^{n_m} \times \mathbb{R}^{n_s}$ with $M \subset B$ such that the trajectories with initial conditions in B tend to M as $t \rightarrow \infty$.

From Definition 7.7 we can say that FGMS is achieved when $\lim_{t\rightarrow\infty} H_{ms}(X_s) - X_m = 0$.

Theorem 7.2. *Let a set of systems as (7.9) and (7.10) be transformable to a MFGOCF. Then* $\lim_{t\rightarrow\infty} \xi_m - \xi_s = 0$, *where* ξ_m *and* ξ_s *are the trajectories in the transformed space of the family of master and slave systems, respectively.*

Proof. Without loss of generality we consider that $u_{m_\mu} = 0$. The set of master systems has the following family of outputs:

$$y_{m_j} = \sum_{i=n-n_j+1}^{n} \gamma_i x_{i,m_j} = \xi_i^{n_{m_j}}$$

where γ_i, $(i = l)$ are differential quantities of u and their time finite derivatives, the family of outputs for slave systems is:

$$y_{s_j} = \sum_{i=n-n_j+1}^{n} \gamma_i x_{i,s_j} + \sum_k \beta_k u_{k,m_j} = \xi_i^{n_{s_j}},$$

where γ_i, $(i = l)$, β_k are differential quantities of u and their time finite derivatives. Then we obtain:

$$\begin{aligned}
\mathscr{D}^\alpha \xi_i^{n_{m_1}} &= \xi_{i+1}^{n_{m_1}}, \quad 1 \le i \le n_{m_1} - 1 \\
\mathscr{D}^\alpha \xi_{n_{m_1}}^{n_{m_1}} &= -\mathscr{L}_{m_1}(\xi_1^{n_{m_1}}, \ldots, \xi_{n_{m_1}}^{n_{m_1}}) \\
\mathscr{D}^\alpha \xi_i^{n_{m_2}} &= \xi_{i+1}^{n_{m_2}}, \quad n_{m_1} + 1 \le i \le n_{m_1} + n_{m_2} - 1 \\
\mathscr{D}^\alpha \xi_{n_{m_1}+n_{m_2}}^{n_{m_2}} &= -\mathscr{L}_{m_2}(\xi_{n_{m_1}+1}^{n_{m_2}}, \ldots, \xi_{n_{m_1}+n_{m_2}}^{n_{m_2}}) \\
&\vdots \\
\mathscr{D}^\alpha \xi_i^{n_{mp}} &= \xi_{i+1}^{n_{mp}}, \quad n_{m_1} + \cdots + n_{m_{p-1}} + 1 \le i \le n_{m_1} + \cdots + n_{m_p} - 1 \\
\mathscr{D}^\alpha \xi_{n_{m_1}+\cdots+n_{m_p}}^{n_{mp}} &= -\mathscr{L}_{m_p}(\xi_{n_{m_1}+\cdots+n_{m_{p-1}}+1}^{n_{mp}}, \ldots, \xi_{n_{m_1}+\cdots+n_{m_p}}^{n_{mp}}),
\end{aligned} \tag{7.11}$$

in a compact form (7.11) can be expressed as:

$$\mathscr{D}^\alpha \xi_m = \mathscr{A} \xi_m - \Phi_m(\mathscr{L}_{m_1}, \ldots, \mathscr{L}_{m_p}) \tag{7.12}$$

Now, let us define the following extended system that represents the family of slave systems and a chain of fractional integrators given by a family of dynamical feedbacks:

$$
\begin{aligned}
\mathscr{D}^{\alpha}\xi_i^{ns1} &= \xi_{i+1}^{ns1}, \quad 1 \le i \le n_{s1}-1\\
\mathscr{D}^{\alpha}\xi_{ns_1}^{ns1} &= -\mathscr{L}_{s1}(\xi_1^{ns1},\dots,\xi_{ns_1}^{ns1},u_{s1},u_{s1}^{(\alpha)},\dots,\mathscr{D}^{[\gamma_{s1}-1]\alpha}u_{s1}) + \mathscr{D}^{\gamma_{s1}\alpha}u_{s1}\\
\mathscr{D}^{\alpha}u_i^{ns1} &= u_{i+1}^{ns1}, \quad 1 \le i \le \gamma_{s1}-1\\
\mathscr{D}^{\alpha}u_{\gamma_{s1}}^{ns1} &= -\mathscr{L}_{m1}(\xi_1^{nm1},\dots,\xi_{nm_1}^{nm1}) + \mathscr{L}_{s1}(\xi_1^{ns1},\dots,\xi_{ns_1}^{ns1},u_{s1},u_{s1}^{(\alpha)},\dots,\mathscr{D}^{[\gamma_{s1}-1]\alpha}u_{s1})\\
&\quad + K_1(\xi^{nm1}-\xi^{ns1})\\
\mathscr{D}^{\alpha}\xi_i^{ns2} &= \xi_{i+1}^{ns2}, \quad n_{s1}+1 \le i \le n_{s1}+n_{s2}-1\\
\mathscr{D}^{\alpha}\xi_{ns_1+ns_2}^{ns2} &= -\mathscr{L}_{s2}(\xi_{ns_1+1}^{ns2},\dots,\xi_{ns_1+ns_2}^{ns2},u_{s2},u_{s2}^{(\alpha)},\dots,\mathscr{D}^{[\gamma_{s2}-1]\alpha}u_{s2}) + \mathscr{D}^{\gamma_{s2}\alpha}u_{s2}\\
\mathscr{D}^{\alpha}u_i^{ns2} &= u_{i+1}^{ns2}, \quad \gamma_{s1}+1 \le i \le \gamma_{s1}+\gamma_{s2}-1\\
\mathscr{D}^{\alpha}u_{\gamma_{s1}+\gamma_{s2}}^{ns2} &= -\mathscr{L}_{m2}(\xi_{nm_1+1}^{nm2},\dots,\xi_{nm_1+nm_2}^{nm2})\\
&\quad + \mathscr{L}_{s2}(\xi_{ns_1+1}^{ns2},\dots,\xi_{ns_1+ns_2}^{ns2},u_{s2},u_{s2}^{(\alpha)},\dots,\mathscr{D}^{[\gamma_{s2}-1]\alpha}u_{s2})\\
&\quad + K_2(\xi^{nm2}-\xi^{ns2})\\
&\ \ \vdots\\
\mathscr{D}^{\alpha}\xi_i^{nsp} &= \xi_{i+1}^{nsp}, \quad n_{s1}+\dots+n_{sp-1}+1 \le i \le n_{s1}+\dots+n_{sp}-1\\
\mathscr{D}^{\alpha}\xi_{ns_1+\dots+ns_p}^{nsp} &= -\mathscr{L}_{sp}(\xi_{ns_1+\dots+ns_{p-1}+1}^{nsp},\dots,\xi_{ns_1+\dots+ns_p}^{nsp},u_{sp},u_{sp}^{(\alpha)},\dots,\mathscr{D}^{[\gamma_{sp}-1]\alpha}u_{sp}) + \mathscr{D}^{\gamma_{sp}\alpha}u_{sp}\\
\mathscr{D}^{\alpha}u_i^{nsp} &= u_{i+1}^{nsp}, \quad \gamma_{s1}+\dots+\gamma_{sp-1}+1 \le i \le \gamma_{s1}+\dots+\gamma_{sp}-1\\
\mathscr{D}^{\alpha}u_{\gamma_{s1}+\dots+\gamma_{sp}}^{nsp} &= -\mathscr{L}_{mp}(\xi^{nmp}) + \mathscr{L}_{sp}(\xi^{nsp},u_{sp},u_{sp}^{(\alpha)},\dots,\mathscr{D}^{[\gamma_{sp}-1]\alpha}u_{sp}) + K_p(\xi^{nmp}-\xi^{nsp})
\end{aligned}
\tag{7.13}
$$

Rewriting (7.13) in a compact form

$$
\begin{aligned}
\mathscr{D}^{\alpha}\xi_s &= \mathscr{A}\xi_s - \Phi_s(\mathscr{L}_{s_1},\dots,\mathscr{L}_{s_p}) + \bar{\mathscr{U}}(\mathscr{D}^{\gamma_{s_1}\alpha}u_{s_1},\dots,\mathscr{D}^{\gamma_{s_p}\alpha}u_{s_p})\\
\mathscr{D}^{\alpha}\mathscr{U} &= \mathscr{M}\mathscr{U} + \bar{\mathscr{U}}\\
\bar{\mathscr{U}} &= \mathscr{K}(\xi_m - \xi_s) - \Phi_m(\mathscr{L}_{m_1},\dots,\mathscr{L}_{m_p}) + \Phi_s(\mathscr{L}_{s_1},\dots,\mathscr{L}_{s_p})
\end{aligned}
$$

where

$$
\mathscr{U} = \begin{bmatrix} u_{n_{s_1}} \\ \vdots \\ u_{n_{s_p}} \end{bmatrix}, \quad u_{n_{s_j}} = \begin{bmatrix} u_1^{n_{s_j}} \\ u_2^{n_{s_j}} \\ \vdots \\ u_{\sum_{j=1}^{j}\gamma_{s_j}}^{n_{s_j}} \end{bmatrix}.
$$

Assume the control signals as $u_1^{n_{s_j}} = u_{s_j}, u_2^{n_{s_j}} = \mathscr{D}^{\alpha} u_{s_j}, \ldots, u_{\alpha_{s_j}}^{n_{s_j}} = \mathscr{D}^{[\gamma_{s_j}-1]\alpha} u_{s_j}$, $\xi^{n_{s_j}} = [\xi^{n_{s_j}}_{n_{s_1}+\cdots+n_{s_{j-1}}+1}, \ldots, \xi^{n_{s_j}}_{n_{s_1}+\cdots+n_{s_j}}]^T, \xi^{n_{m_j}} = [\xi^{n_{m_j}}_{n_{m_1}+\cdots+n_{m_{j-1}}+1}, \ldots, \xi^{n_{m_j}}_{n_{m_1}+\cdots+n_{m_j}}]^T$ and $K_j = [k_{1,j}, \ldots, k_{n_j,j}]$. Matrices $\mathscr{M}$ and $\mathscr{K}$ are defined as follows:

$$\mathscr{M} = \begin{bmatrix} \bar{\mathscr{M}_1} & & 0 \\ & \ddots & \\ 0 & & \bar{\mathscr{M}_p} \end{bmatrix}, \quad \bar{\mathscr{M}_j} = \begin{bmatrix} 0\,1\,0 & 0 & \ldots & 0 \\ 0\,0\,1 & 0 & \ldots & 0 \\ \vdots\;\vdots\;\vdots & \ddots & \ldots & 0 \\ 0\,0\,0 & 0 & 1 & 0 \\ 0\,0\,0 & 0 & 0 & 1 \\ 0\,0\,0 & 0 & 0 & 0 \end{bmatrix},$$

$$\mathscr{K} = \begin{bmatrix} \bar{\mathscr{K}_1} & & 0 \\ & \ddots & \\ 0 & & \bar{\mathscr{K}_p} \end{bmatrix}, \quad \bar{\mathscr{K}_j} = \begin{bmatrix} 0 & 0 & 0 & 0 & \ldots & 0 \\ 0 & 0 & 0 & 0 & \ldots & 0 \\ \vdots & \vdots & \vdots & \vdots & \ldots & \vdots \\ 0 & 0 & 0 & 0 & \ldots & 0 \\ 0 & 0 & 0 & 0 & \ldots & 0 \\ k_{1,j} & k_{2,j} & k_{3,j} & k_{4,j} & \ldots & k_{n_j,j} \end{bmatrix}$$

Finally we consider the error of synchronization $e_{\xi} = \xi_m - \xi_s$ that has a dynamics given by:

$$\begin{aligned} \mathscr{D}^{\alpha} e_{\xi} &= \mathscr{A}\xi_m - \Phi_m(\mathscr{L}_{m_1}, \ldots, \mathscr{L}_{m_p}) - \mathscr{A}\xi_s + \Phi_s(\mathscr{L}_{s_1}, \ldots, \mathscr{L}_{s_p}) - \bar{\mathscr{U}} \\ \mathscr{D}^{\alpha}\mathscr{U} &= \mathscr{M}\mathscr{U} + \bar{\mathscr{U}} \\ \bar{\mathscr{U}} &= \mathscr{K}(\xi_m - \xi_s) - \Phi_m(\mathscr{L}_{m_1}, \ldots, \mathscr{L}_{m_p}) + \Phi_s(\mathscr{L}_{s_1}, \ldots, \mathscr{L}_{s_p}) \end{aligned}$$

and after some algebraic manipulations

$$\mathscr{D}^{(\alpha)} e_{\xi} = (\mathscr{A} - \mathscr{K})e_{\xi} \tag{7.14}$$

from Theorem 7.1, the system (7.14) is asymptotically stable if all eigenvalues of matrix $\mathscr{A} - \mathscr{K} = diag(\bar{\mathscr{A}_1}, \ldots, \bar{\mathscr{A}_p})$ with control gains $(k_{1,j}, k_{2,j}, \ldots, k_{n_j,j})$ are chosen such that:

$$|arg(\lambda_i(\bar{\mathscr{A}_j}))| > \frac{\pi}{2} > \alpha\frac{\pi}{2}$$

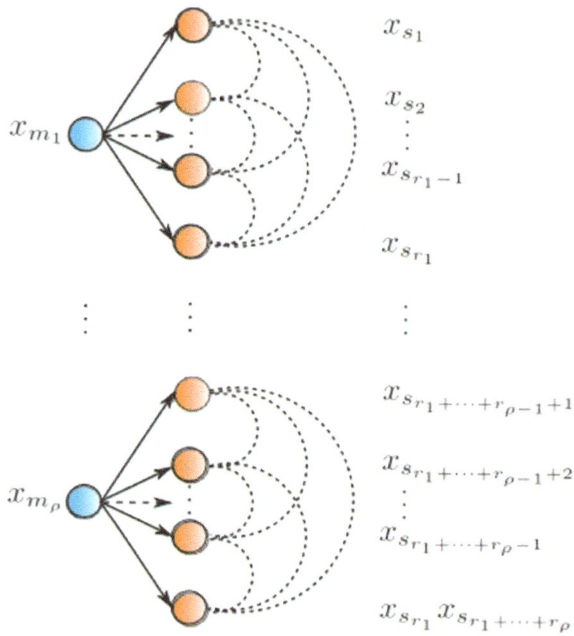

Fig. 7.2 Slave interactions in complex network

where

$$\bar{\mathscr{A}_j} = \begin{bmatrix} 0 & 1 & 0 & 0 & \dots & 0 \\ 0 & 0 & 1 & 0 & \dots & 0 \\ \vdots & \vdots & \vdots & \ddots & \dots & 0 \\ 0 & 0 & 0 & 0 & 1 & 0 \\ 0 & 0 & 0 & 0 & 0 & 1 \\ -k_{1,j} & -k_{2,j} & -k_{3,j} & -k_{4,j} & \dots & -k_{n_j,j} \end{bmatrix}$$

and the proof is completed (Fig. 7.2). □

Corollary 7.1. *A family of Liouvillian fractional-order system class that is a family of PV systems is in a state of FGMS.*

Proof. The proof is trivial and it is omitted (transitivity). □

7.2.1 Numerical Examples of FGMS

In this section, it is shown the effectiveness of the proposed methodology. Moreover, it is shown that is not restricted to Liouvillian type systems. This is the case given in the last two networks involving Rössler systems, which are not Liouvillian systems. The first example considers the case of GS. The second example considers the case of FGMS with one master system where CS and GS are achieved. Third example consists of the case of two families of slave systems without interaction.

Example 7.1. Let a master–slave configuration with Arneodo and Chua-Hartley systems as master and slave, respectively. The objective is to achieve the state of GS and show the Liouvillian feature of these systems.

First, let the master system be:

$$\begin{aligned}
\mathscr{D}^{\alpha} x_1^{m_1} &= x_2^{m_1} \\
\mathscr{D}^{\alpha} x_2^{m_1} &= x_3^{m_1} \\
\mathscr{D}^{\alpha} x_3^{m_1} &= -\beta_1 x_1^{m_1} - \beta_2 x_2^{m_1} - b_3 x_3^{m_1} + \beta_4 x_1^{m_1 3}
\end{aligned} \tag{7.15}$$

assume $y_{m_1} = x_2^{m_1}$ as output, such that the states of system 7.15 can be represented as a function of the output:

$$\begin{aligned}
\mathscr{D}^{\alpha} x_1^{m_1} &= y_{m_1} \\
x_2^{m_1} &= y_{m_1} \\
x_3 &= y_{m_1}^{(\alpha)}
\end{aligned}$$

Hence, state $x_1^{m_1}$ satisfies FLAO condition. Now, note that $x_1^{m_1}$ can be written as a function of a fractional integral of $x_2^{m_1}$, that is to say:

$$\begin{aligned}
x_1^{m_1} &= I^{\alpha} y_{m_1} \\
x_2^{m_1} &= y_{m_1} \\
x_3^{m_1} &= y_{m_1}^{(\alpha)}
\end{aligned}$$

thus, system (7.15) is a fractional-order Liouvillian system. On the other hand, let us verify the observability condition that slave system fulfills. Let slave system be:

$$\begin{aligned}
\mathscr{D}^{\alpha} x_1^{s_1} &= \rho \left(x_2^{s_1} + \frac{x_1^{s_1} - 2 x_1^{s_1 3}}{7} \right) \\
\mathscr{D}^{\alpha} x_2^{s_1} &= x_1^{s_1} - x_2^{s_1} + x_3^{s_1} \\
\mathscr{D}^{\alpha} x_3^{s_1} &= -\beta x_2^{s_1}
\end{aligned} \tag{7.16}$$

Assume $y_{s_1} = x_2^{s_1}$ as output, such that the states of system (7.16) can be represented as a function of the output, i.e.,

$$
\begin{aligned}
x_1^{s_1} &= y_{s_1} + y_{s_1}^{(\alpha)} - x_3^{s_1} \\
x_2^{s_1} &= y_{s_1} \\
\mathscr{D}^\alpha x_3^{s_1} &= -\beta y_{s_1}
\end{aligned}
$$

Hence, states $x_1^{s_1}$ and $x_3^{s_1}$ satisfy FLAO condition. Note that we can choose $\bar{y}_{s_1} = I^\alpha y_{s_1} + u_1^{s_1}$ such that $x_1^{s_1}$ and $x_3^{s_1}$ can be obtained as a function of fractional integrals of $x_2^{s_1}$, that is to say:

$$
\begin{aligned}
x_1^{s_1} &= y_{s_1} + y_{s_1}^{(\alpha)} + \beta I^\alpha y_{s_1} - u_1^{s_1} \\
x_2^{s_1} &= y_{s_1} \\
x_3^{s_1} &= -\beta I^\alpha y_{s_1} + u_1^{s_1}
\end{aligned}
$$

The system (7.16) is a Fractional Liouvillian system. Now assume systems (7.15) and (7.16). Consider the family of outputs for the master and slave systems, respectively, as follows:

$$
\bar{y}_{m_1} = I^\alpha y_{m_1} = x_1^{m_1}
$$

and

$$
\bar{y}_{s_1} = I^\alpha y_{s_1} + u_1^{s_1} = -\frac{1}{\beta} x_3^{s_1} + u_1^{s_1}
$$

From the family of outputs next transformations are fulfilled for the master system:

$$
\xi_m = \begin{pmatrix} \xi_1^{m_1} \\ \xi_2^{m_1} \\ \xi_3^{m_1} \end{pmatrix} = \begin{pmatrix} I^\alpha y_{m_1} \\ y_{m_1} \\ \mathscr{D}^\alpha y_{m_1} \end{pmatrix} = \begin{pmatrix} x_1^{m_1} \\ x_2^{m_1} \\ x_3^{m_1} \end{pmatrix} = \Phi_m(X_m)
$$

with its inverse:

$$
X_m = \begin{pmatrix} x_1^{m_1} \\ x_2^{m_1} \\ x_3^{m_3} \end{pmatrix} = \begin{pmatrix} \xi_1^{m_1} \\ \xi_2^{m_1} \\ \xi_3^{m_1} \end{pmatrix} = \Phi_m^{-1}(\xi_m)
$$

For the slave system, it is obtained

$$
\xi_s = \begin{pmatrix} \xi_1^{s_1} \\ \xi_2^{s_1} \\ \xi_3^{s_1} \end{pmatrix} = \begin{pmatrix} I^\alpha y_{s_1} + u_1^{s_1} \\ y_{s_1} + u_2^{s_1} \\ \mathscr{D}^\alpha y_{s_1} + u_3^{s_1} \end{pmatrix} = \begin{pmatrix} -\frac{1}{\beta} x_3^{s_1} + u_1^{s_1} \\ x_2^{s_1} + u_2^{s_1} \\ x_1^{s_1} - x_2^{s_1} + x_3^{s_1} + u_3^{s_1} \end{pmatrix} = \Phi_s(X_s)
$$

with its inverse:

$$X_s = \begin{pmatrix} x_1^{s_1} \\ x_2^{s_1} \\ x_3^{s_1} \end{pmatrix} = \begin{pmatrix} \beta(\xi_1^{s_1} - u_1^{s_1}) + \xi_2^{s_1} - u_2^{s_1} + \xi_3^{s_1} - u_3^{s_1} \\ \xi_2^{s_1} - u_2^{s_1} \\ -\beta(\xi_1^{s_1} - u_1^{s_1}) \end{pmatrix} = \Phi_s^{-1}(\xi_s)$$

Then, master in transformed coordinates is given by:

$$\begin{aligned} \mathscr{D}^\alpha \xi_1^{m_1} &= \xi_2^{m_1} \\ \mathscr{D}^\alpha \xi_2^{m_1} &= \xi_3^{m_1} \\ \mathscr{D}^\alpha \xi_3^{m_1} &= -\mathscr{L}_{m_1}(\xi_1^{m_1}, \xi_2^{m_1}, \xi_3^{m_1}) \end{aligned}$$

with

$$\mathscr{L}_{m_1}(\cdot) = \beta_1(\xi_1^{s_1} - u_1^{s_1}) + \beta_2(\xi_2^{s_1} - u_2^{s_1}) + \beta_3(\xi_3^{s_1} - u_3^{s_1}) - \beta_4(\xi_1^{s_1} - u_1^{s_1})^3$$

and the slave system in transformed coordinates is given by:

$$\begin{aligned} \mathscr{D}^\alpha \xi_1^{s_1} &= \xi_2^{s_1} \\ \mathscr{D}^\alpha \xi_2^{s_1} &= \xi_3^{s_1} \\ \mathscr{D}^\alpha \xi_3^{s_1} &= -\mathscr{L}_{s_1}(\xi_1^{s_1}, \xi_2^{s_1}, \xi_3^{s_1}, u_1^{s_1}, u_2^{s_1}, u_3^{s_1}) + \bar{u}_{s_1} \\ \mathscr{D}^\alpha u_1^{s_1} &= u_2^{s_1} \\ \mathscr{D}^\alpha u_2^{s_1} &= u_3^{s_1} \\ \mathscr{D}^\alpha u_3^{s_1} &= \bar{u}_{s_1} \end{aligned}$$

with

$$\begin{aligned} \mathscr{L}_{s_1}(\cdot) = &-\frac{\rho\beta}{7}(\xi_1^{s_1} - u_1^{s_1}) - \left(\frac{8\rho - 7\beta}{7}\right)(\xi_2^{s_1} - u_2^{s_1}) - \left(\frac{\rho - 7}{7}\right)(\xi_3^{s_1} - u_3^{s_1}) \\ &+ \frac{7\rho}{7}\left((\xi_1^{s_1} - u_1^{s_1}) + (\xi_2^{s_1} - u_2^{s_1}) + (\xi_3^{s_1} - u_3^{s_1})\right)^3 \end{aligned}$$

The closed-loop dynamics for synchronization error $e_\xi = \xi_m - \xi_s$ is represented in the next augmented system:

$$\begin{aligned} \mathscr{D}^\alpha e_{\xi_1}^{s_1} &= e_{\xi_2}^{s_1} \\ \mathscr{D}^\alpha e_{\xi_2}^{s_1} &= e_{\xi_3}^{s_1} \\ \mathscr{D}^\alpha e_{\xi_3}^{s_1} &= -\mathscr{L}_{m_1}(\xi_1^{m_1}, \xi_2^{m_1}, \xi_3^{m_1}) + \mathscr{L}_{s_1}(\xi_1^{s_1}, \xi_2^{s_1}, \xi_3^{s_1}, u_1^{s_1}, u_2^{s_1}, u_3^{s_1}) - \mathscr{D}^\alpha u_3^{s_1} \\ \mathscr{D}^\alpha u_1^{s_1} &= u_2^{s_1} \\ \mathscr{D}^\alpha u_2^{s_1} &= u_3^{s_1} \\ \mathscr{D}^\alpha u_3^{s_1} &= -\mathscr{L}_{m_1}(\xi_1^{m_1}, \xi_2^{m_1}, \xi_3^{m_1}) + \mathscr{L}_{s_1}(\xi_1^{s_1}, \xi_2^{s_1}, \xi_3^{s_1}, u_1^{s_1}, u_2^{s_1}, u_3^{s_1}) + k_{s_1} e_\xi^{s_1} \end{aligned}$$

that is, $\mathscr{D}^{\alpha} e_{\xi} = (\mathscr{A} - \mathscr{K}) e_{\xi}$. Then synchronization error converges asymptotically to zero if matrix $\mathscr{A} - \mathscr{K} = \bar{\mathscr{A}_1}$ is Hurwitz. Where

$$\bar{\mathscr{A}_1} = \begin{bmatrix} 0 & 1 & 0 \\ 0 & 0 & 0 \\ -k_{1,1} & -k_{2,1} & -k_{3,1} \end{bmatrix}$$

This is given when $k_{1,1}, k_{1,3} > 0$ and $k_{1,2} > k_{1,1}/k_{1,3}$. Parameters for master and slave systems are $\rho = 12.75$, $\beta = 100/7$, $\beta_1 = -5.5$, $\beta_2 = 3.5$, $\beta_3 = 0.8$, $\beta_4 = -1$, commensurate fractional order $\alpha = 0.92$, initial conditions $x_{m_1}(0) = \left[-0.20\ 0.35\ 0.20\right]^T$, $x_{s_1}(0) = \left[-0.58\ -0.01\ 0.30\right]^T$ to obtain a chaotic behavior and $k_{s,1} = \left[10\ 10\ 10\right]$.

In Figs. 7.3 and 7.4, GS is shown in original and transformed coordinates. This case illustrates how GS is given in a master–slave configuration. Figures 7.3 and 7.4 apparently are the same due to master system being given in a canonical form (transformation $\Phi_m(\cdot)$ is equal to the identity), that makes the mapping

$$H_{ms}(X_s) = \Phi_m^{-1} \circ \Phi_s(X_s) = \begin{pmatrix} -\frac{1}{\beta} x_3^{s_1} + u_1^{s_1} \\ x_2^{s_1} + u_2^{s_1} \\ x_1^{s_1} - x_2^{s_1} + x_3^{s_1} + u_3^{s_1} \end{pmatrix} \tag{7.17}$$

to be easily obtained. The structure of the mapping is directly obtained from $\Phi_s(\cdot)$, which means that synchronized original and transformed coordinates are equal; therefore we obtain similar trajectories. Original trajectories for the slave system are obtained from mapping (7.17). Synchronization error asymptotically converges to the origin as seen in Figs. 7.3a and 7.4a. Finally, note that GS trajectories converge to the same chaotic attractor. Convergence can be verified from time evolution of the synchronized trajectories (see Figs. 7.3b, c and 7.4b, c).

Example 7.2. Consider two strictly different commensurate fractional-order Liouvillian systems. The first one is a Chua-Hartley system as master, and let Arneodo and Chua-Hartley as slave systems. This configuration is illustrated in Fig. 7.5. The objective is to achieve the state of CS and GS for these Liouvillian systems, hence the state of FGMS.

Consider the master system (Chua-Hartley) given by

$$\begin{aligned}
\mathscr{D}^{\alpha} x_1^{m_1} &= \rho \left(x_2^{m_1} + \frac{x_1^{m_1} - 2 x_1^{m_1 3}}{7} \right) \\
\mathscr{D}^{\alpha} x_2^{m_1} &= x_1^{m_1} - x_2^{m_1} + x_3^{m_1} \\
\mathscr{D}^{\alpha} x_3^{m_1} &= -\beta x_2^{m_1}
\end{aligned} \tag{7.18}$$

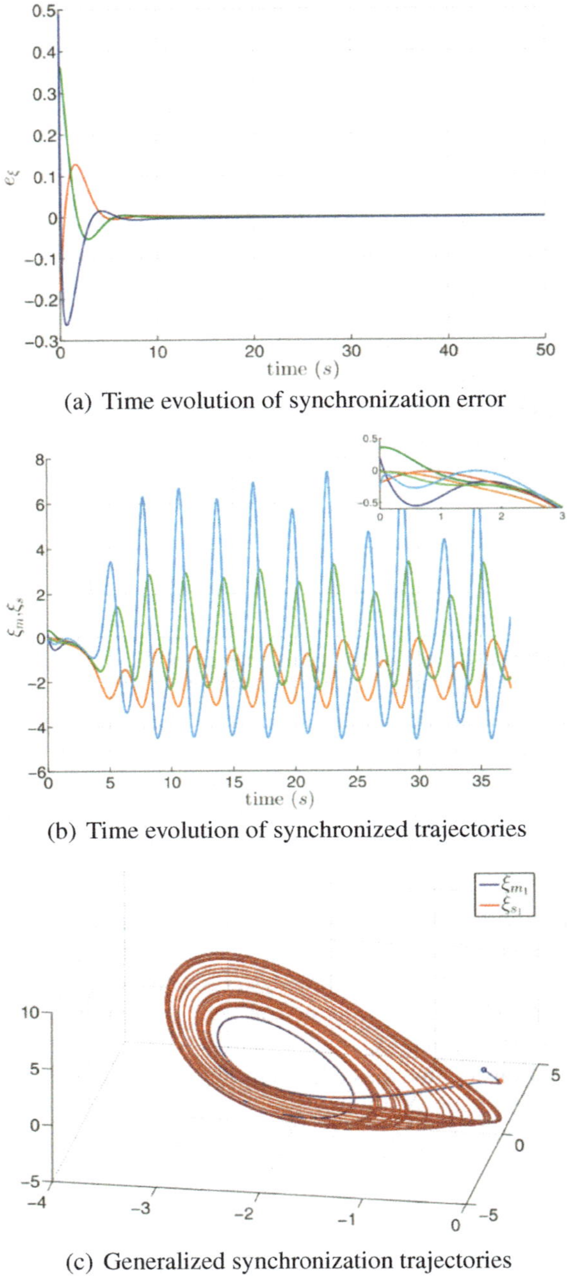

(a) Time evolution of synchronization error

(b) Time evolution of synchronized trajectories

(c) Generalized synchronization trajectories

Fig. 7.3 GS in transformed coordinates (identical systems)

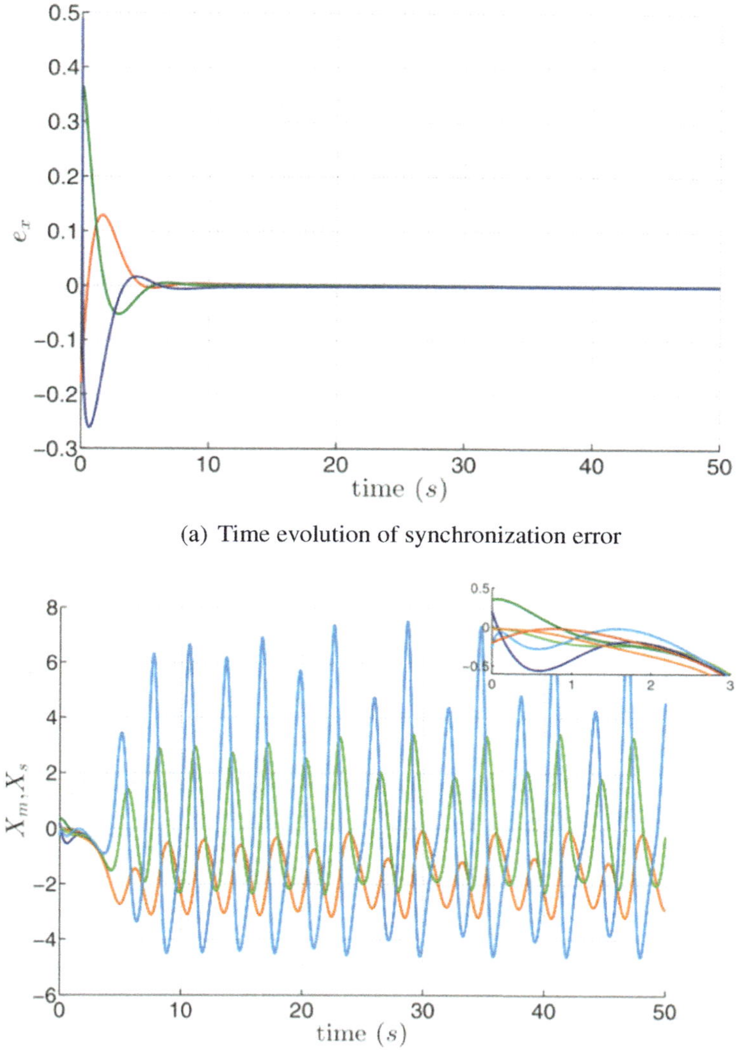

(a) Time evolution of synchronization error

(b) Time evolution of synchronized trajectories

Fig. 7.4 GS in original coordinates (identical systems)

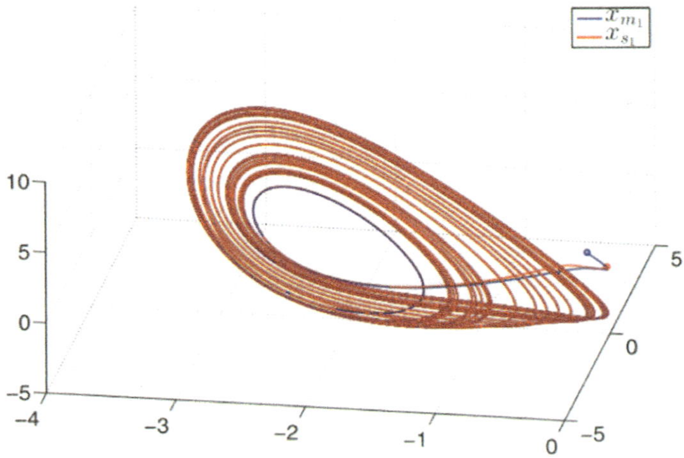

(c) Generalized synchronization trajectories

Fig. 7.4 (continued)

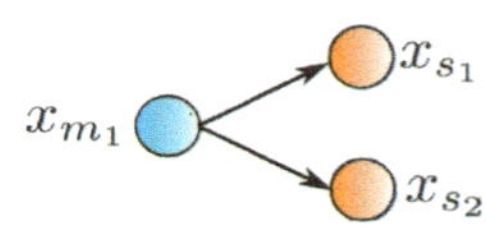

Fig. 7.5 Configuration of master system x_{m_1} and slave systems x_{s_1}, x_{s_2}

Assume $y_{m_1} = x_2^{m_1}$ as output, then the states of system (7.18) can be expressed as a function of the output, i.e.,

$$\begin{aligned} x_1^{m_1} &= y_{m_1} + y_{m_1}^{(\alpha)} - x_3^{m_1} \\ x_2^{m_1} &= y_{m_1} \\ \mathscr{D}^{\alpha} x_3^{m_1} &= -\beta y_{m_1} \end{aligned}$$

Hence, states $x_1^{m_1}$ and $x_3^{m_1}$ satisfy FLAO condition. On the other hand, rewrite $x_1^{m_1}$ and $x_3^{m_1}$ as a function of fractional integrals of $x_2^{m_1}$, that is to say:

$$\begin{aligned} x_1^{m_1} &= y_{m_1} + y_{m_1}^{(\alpha)} + \beta I^{\alpha} y_{m_1} \\ x_2^{m_1} &= y_{m_1} \\ x_3^{m_1} &= -\beta I^{(\alpha)} y_{m_1} \end{aligned}$$

system (7.18) is a Fractional Liouvillian system.

Now, consider the master system (7.18) and assume dynamics of the family of slave systems as

$$
\begin{aligned}
\mathscr{D}^{\alpha} x_1^{s_1} &= x_2^{s_1} \\
\mathscr{D}^{\alpha} x_2^{s_1} &= x_3^{s_1} \\
\mathscr{D}^{\alpha} x_3^{s_1} &= -\beta_1 x_1^{s_1} - \beta_2 x_2^{s_1} - \beta_3 x_3^{s_1} + \beta_4 x_1^{s_1 3} \\
\mathscr{D}^{\alpha} x_4^{s_2} &= \rho \left(x_5^{s_2} + \frac{x_4^{s_2} - 2x_4^{s_2 3}}{7} \right) \\
\mathscr{D}^{\alpha} x_5^{s_2} &= x_4^{s_2} - x_5^{s_2} + x_6^{s_2} \\
\mathscr{D}^{\alpha} x_6^{s_2} &= -\beta x_5^{s_2}
\end{aligned}
$$

Consider the family of outputs for the family of master and slave systems, respectively, as follows

$$
\bar{y}_{m_1} = I^{\alpha} y_{m_1} = -\frac{1}{\beta} x_3^{m_1}
$$

and

$$
\begin{aligned}
\bar{y}_{s_1} &= I^{\alpha} y_{s_1} + u_1^{s_1} = x_1^{s_1} + u_1^{s_1} \\
\bar{y}_{s_2} &= I^{\alpha} y_{s_2} + u_4^{s_2} = -\frac{1}{\beta} x_6^{s_2} + u_4^{s_2}
\end{aligned}
$$

Then, master in transformed coordinates is given by

$$
\begin{aligned}
\mathscr{D}^{\alpha} \xi_1^{m_1} &= \xi_2^{m_1} \\
\mathscr{D}^{\alpha} \xi_2^{m_1} &= \xi_3^{m_1} \\
\mathscr{D}^{\alpha} \xi_3^{m_1} &= -\mathscr{L}_{m_1}(\xi_1^{m_1}, \xi_2^{m_1}, \xi_3^{m_1})
\end{aligned}
$$

with

$$
\mathscr{L}_{m_1}(\cdot) = -\frac{\rho\beta}{7} \xi_1^{m_1} - \left(\frac{8\rho - 7\beta}{7} \right) \xi_2^{m_1} - \left(\frac{\rho - 7}{7} \right) \xi_3^{m_1} + \frac{2\rho}{7} (\xi_1^{m_1} + \xi_2^{m_1} + \xi_3^{m_1})^3
$$

and the family of slave systems in transformed coordinates is given by

$$
\begin{aligned}
\mathscr{D}^{\alpha} \xi_1^{s_1} &= \xi_2^{s_1} \\
\mathscr{D}^{\alpha} \xi_2^{s_1} &= \xi_3^{s_1} \\
\mathscr{D}^{\alpha} \xi_3^{s_1} &= -\mathscr{L}_{s_1}(\xi_1^{s_1}, \xi_2^{s_1}, \xi_3^{s_1}, u_1^{s_1}, u_2^{s_1}, u_3^{s_1}) + \bar{u}_{s_1} \\
\mathscr{D}^{\alpha} u_1^{s_1} &= u_2^{s_1} \\
\mathscr{D}^{\alpha} u_2^{s_1} &= u_3^{s_1}
\end{aligned}
$$

$$
\begin{aligned}
\mathscr{D}^{\alpha} u_3^{s_1} &= \bar{u}_{s_1} \\
\mathscr{D}^{\alpha} \xi_4^{s_2} &= \xi_5^{s_2} \\
\mathscr{D}^{\alpha} \xi_5^{s_2} &= \xi_6^{s_2} \\
\mathscr{D}^{\alpha} \xi_6^{s_2} &= -\mathscr{L}_{s_2}(\xi_4^{s_2}, \xi_5^{s_2}, \xi_6^{s_2}, u_4^{s_2}, u_5^{s_2}, u_6^{s_2}) + \bar{u}_{s_2} \\
\mathscr{D}^{\alpha} u_4^{s_2} &= u_5^{s_2} \\
\mathscr{D}^{\alpha} u_5^{s_2} &= u_6^{s_2} \\
\mathscr{D}^{\alpha} u_6^{s_2} &= \bar{u}_{s_2}
\end{aligned}
$$

with

$$
\begin{aligned}
\mathscr{L}_{s_1}(\cdot) &= \beta_1(\xi_1^{s_1} - u_1^{s_1}) + \beta_2(\xi_2^{s_1} - u_2^{s_1}) + \beta_3(\xi_3^{s_1} - u_3^{s_1}) - \beta_4(\xi_1^{s_1} - u_1^{s_1})^3 \\
\mathscr{L}_{s_2}(\cdot) &= -\frac{\rho\beta}{7}(\xi_4^{s_2} - u_4^{s_2}) - \left(\frac{8\rho - 7\beta}{7}\right)(\xi_5^{s_2} - u_5^{s_2}) - \left(\frac{\rho - 7}{7}\right)(\xi_6^{s_2} - u_6^{s_2}) \\
&\quad + \frac{2\rho}{7}\left(\beta(\xi_4^{s_2} - u_4^{s_2}) + (\xi_5^{s_2} - u_5^{s_2}) + (\xi_6^{s_2} - u_6^{s_2})\right)^3
\end{aligned}
$$

Remark 7.2. Notice that the dimension of master systems is extended via virtual master systems that will have the same dynamics and initial conditions as the original master systems associated with the corresponding slave system.

The closed-loop dynamics of synchronization error $e_\xi = \xi_m - \xi_s$ is given by the following augmented system:

$$
\begin{aligned}
\mathscr{D}^{\alpha} e_{\xi_1}^{s_1} &= e_{\xi_2}^{s_1} \\
\mathscr{D}^{\alpha} e_{\xi_2}^{s_1} &= e_{\xi_3}^{s_1} \\
\mathscr{D}^{\alpha} e_{\xi_3}^{s_1} &= -\mathscr{L}_{m_1}(\xi_1^{m_1}, \xi_2^{m_1}, \xi_3^{m_1}) + \mathscr{L}_{s_1}(\xi_1^{s_1}, \xi_2^{s_1}, \xi_3^{s_1}, u_1^{s_1}, u_2^{s_1}, u_3^{s_1}) - \mathscr{D}^{\alpha} u_3^{s_1} \\
\mathscr{D}^{\alpha} u_1^{s_1} &= u_2^{s_1} \\
\mathscr{D}^{\alpha} u_2^{s_1} &= u_3^{s_1} \\
\mathscr{D}^{\alpha} u_3^{s_1} &= -\mathscr{L}_{m_1}(\xi_1^{m_1}, \xi_2^{m_1}, \xi_3^{m_1}) + \mathscr{L}_{s_1}(\xi_1^{s_1}, \xi_2^{s_1}, \xi_3^{s_1}, u_1^{s_1}, u_2^{s_1}, u_3^{s_1}) + k_{s_1} e_\xi^{s_1} \\
\mathscr{D}^{\alpha} e_{\xi_4}^{s_2} &= e_{\xi_5}^{s_2} \\
\mathscr{D}^{\alpha} e_{\xi_5}^{s_2} &= e_{\xi_6}^{s_2} \\
\mathscr{D}^{\alpha} e_{\xi_6}^{s_2} &= -\mathscr{L}_{m_1}(\xi_1^{m_1}, \xi_2^{m_1}, \xi_3^{m_1}) + \mathscr{L}_{s_2}(\xi_4^{s_2}, \xi_5^{s_2}, \xi_6^{s_2}, u_4^{s_2}, u_5^{s_2}, u_6^{s_2}) - \mathscr{D}^{\alpha} u_6^{s_2} \\
\mathscr{D}^{\alpha} u_4^{s_2} &= u_5^{s_2} \\
\mathscr{D}^{\alpha} u_5^{s_2} &= u_6^{s_2} \\
\mathscr{D}^{\alpha} u_6^{s_2} &= -\mathscr{L}_{m_1}(\xi_1^{m_1}, \xi_2^{m_1}, \xi_3^{m_1}) + \mathscr{L}_{s_2}(\xi_4^{s_2}, \xi_5^{s_2}, \xi_6^{s_2}, u_4^{s_2}, u_5^{s_2}, u_6^{s_2}) + k_{s_2} e_\xi^{s_2}
\end{aligned}
$$

After some algebraic manipulations, it is obtained $D^{\alpha} e_\xi = (\mathscr{A} - \mathscr{K}) e_\xi$. Then synchronization error converges asymptotically to zero if matrix $\mathscr{A} - \mathscr{K} = diag$ $(\bar{\mathscr{A}_1}, \bar{\mathscr{A}_2})$ is Hurwitz. Where

$$\bar{\mathscr{A}_j} = \begin{bmatrix} 0 & 1 & 0 \\ 0 & 0 & 1 \\ -k_{1,j} & -k_{2,j} & -k_{3,j} \end{bmatrix}, \quad 1 \leq j \leq 2.$$

Parameters are taken as: Initial conditions $x_{m_1}(0) = \left[-0.50\ -0.07\ 0.65\right]^T$, $x_{s_1}(0) = \left[-0.20\ 0.35\ 0.20\right]^T$, $x_{s_2}(0) = \left[-0.58\ -0.01\ 0.30\right]^T$ to ensure chaotic behavior and $k_{s,j} = \left[200\ 200\ 200\right]$ for $1 \leq j \leq 2$.

Figure 7.6 illustrates transformed coordinates of master and slave systems that are in state of FGMS, Fig. 7.6a shows error synchronization convergence in transformed coordinates, and in Fig. 7.7 are shown the states of master and slave systems and synchronization error convergence in original coordinates with

$$H_{ms}(X_s) = \begin{pmatrix} \beta(x_1^{s_1} + u_1^{s_1}) + x_2^{s_1} + u_2^{s_1} + x_3^{s_1} + u_3^{s_1} \\ x_2^{s_1} + u_2^{s_1} \\ -\beta(x_1^{s_1} + u_1^{s_1}) \\ x_4^{s_2} + \beta u_4^{s_2} + u_5^{s_2} + u_6^{s_2} \\ x_5^{s_2} + u_5^{s_2} \\ -\beta(x_4^{s_2} - x_5^{s_2} + x_6^{s_2} + u_6^{s_2}) \end{pmatrix}$$

The effectiveness of the approach can be verified from the multi-synchronization trajectories. Note that the synchronization error converges asymptotically to the origin.

Example 7.3. Assume the configuration is composed of two master systems given in Fig. 7.8. The objective is to synchronize the decoupled groups of slave systems with their associated master system.

The first group is considered as the system given in example above. Consider the second master as a Rössler system with Arneodo, Chua and Rössler systems as slaves. Let the second master system be

$$\mathscr{D}^\alpha x_4^{m_2} = -(x_5^{m_2} + x_6^{m_2}) \tag{7.19}$$
$$\mathscr{D}^\alpha x_5^{m_2} = x_4^{m_2} + a x_5^{m_2} \tag{7.20}$$
$$\mathscr{D}^\alpha x_6^{m_2} = b + x_6^{m_2}(x_4^{m_2} - c) \tag{7.21}$$

assume $y_{m_2} = x_5^{m_2}$ as output, such that the states of system (7.19) can be represented as a function of the output

$$\begin{aligned} x_4^{m_2} &= \mathscr{D} y_{m_2} + a y_{m_2} \\ x_5^{m_2} &= y_{m_2} \\ x_6^{m_2} &= -\mathscr{D}^{2\alpha} y_{m_1} - \mathscr{D}^\alpha y_{m_1} + y_{m_1} \end{aligned}$$

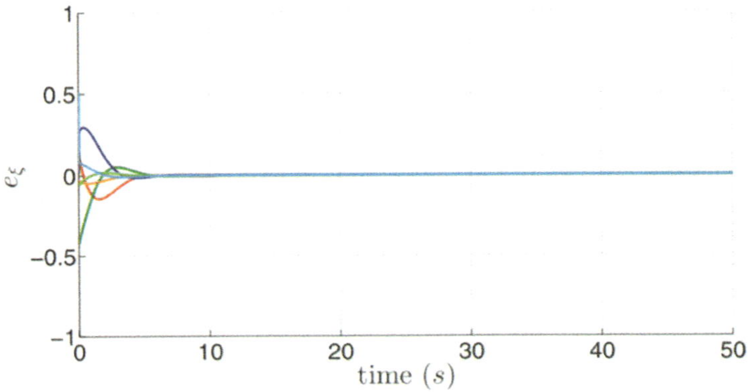

(a) Time evolution of synchronization error

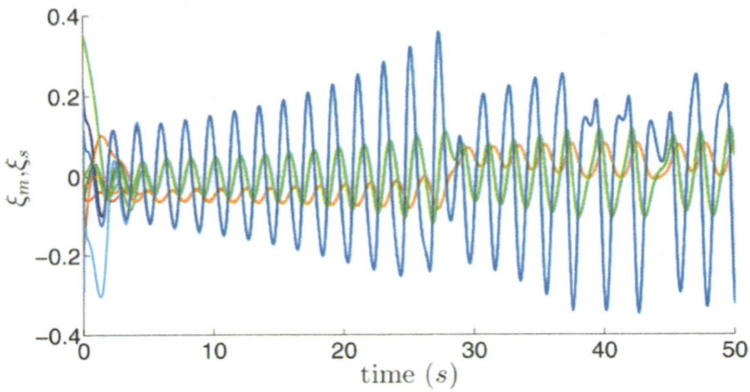

(b) Time evolution of synchronized trajectories

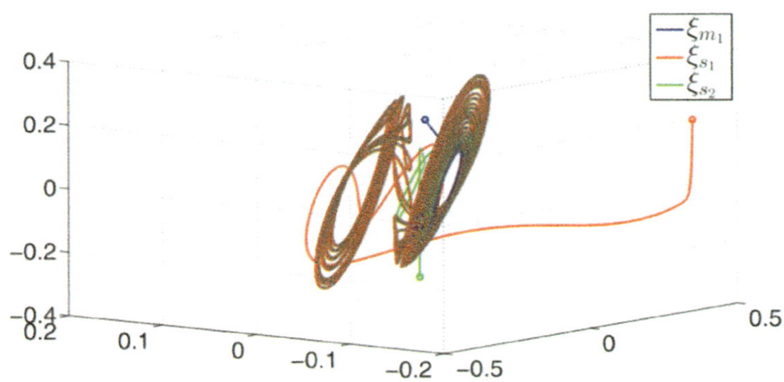

(c) Multi-synchronization trajectories

Fig. 7.6 FGMS in transformed coordinates (different systems)

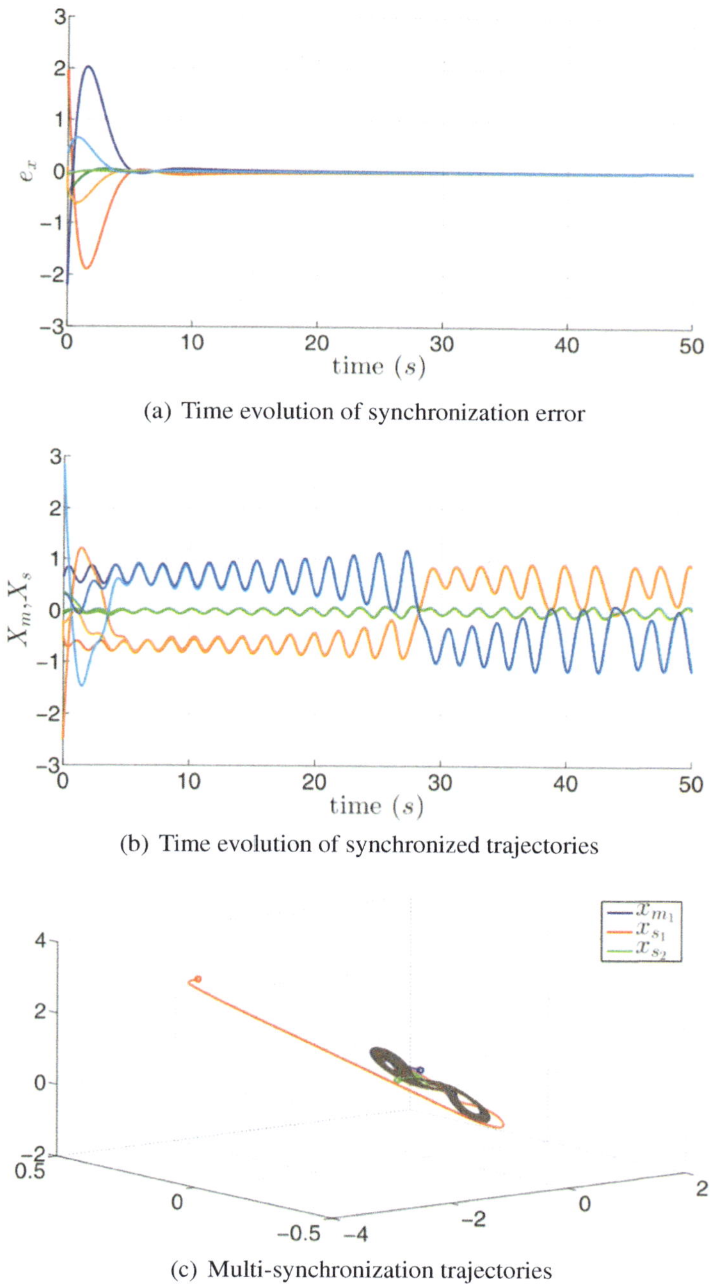

Fig. 7.7 FGMS in original coordinates (different systems)

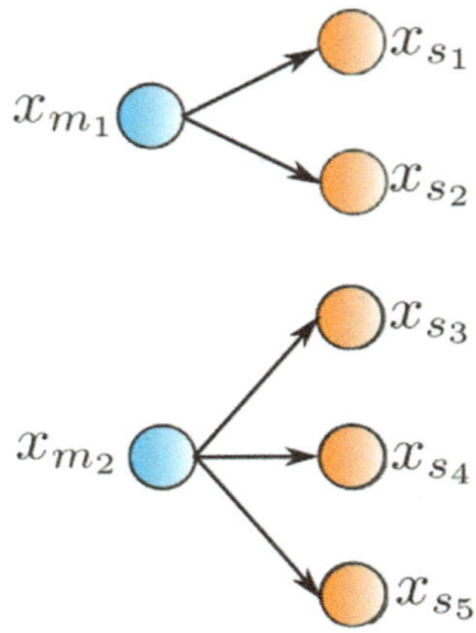

Fig. 7.8 Configuration of master system x_{m_1} and slave systems x_{s_1}, x_{s_2}

Hence the states of system (7.19) satisfy FAO condition. Consider the family of outputs for the family of master and slave systems, respectively, as follows

$$\begin{aligned}\bar{y}_{m_1} &= I^{\alpha} y_{m_1} = -\frac{1}{\beta} x_3^{m_1}\\ \bar{y}_{m_2} &= y_{m_2} = x_5^{m_2}\end{aligned}$$

and

$$\begin{aligned}\bar{y}_{s_1} &= I^{\alpha} y_{s_1} + u_1^{s_1} = x_1^{s_1} + u_1^{s_1}\\ \bar{y}_{s_2} &= I^{\alpha} y_{s_2} + u_4^{s_2} = -\frac{1}{\beta} x_6^{s_2} + u_1^{s_2}\\ \bar{y}_{s_3} &= I^{\alpha} y_{s_3} + u_7^{s_3} = x_7^{s_3} + u_7^{s_3}\\ \bar{y}_{s_4} &= I^{\alpha} y_{s_4} + u_{10}^{s_4} = -\frac{1}{\beta} x_{12}^{s_4} + u_{10}^{s_4}\\ \bar{y}_{s_5} &= y_{s_5} + u_{13}^{s_5} = x_{14}^{s_5} + u_{13}^{s_5}\end{aligned}$$

Then, master in transformed coordinates is given by

$$\begin{aligned}\mathscr{D}^{\alpha}\xi_1^{m_1} &= \xi_2^{m_1}\\ \mathscr{D}^{\alpha}\xi_2^{m_1} &= \xi_3^{m_1}\\ \mathscr{D}^{\alpha}\xi_3^{m_1} &= -\mathscr{L}_{m_1}(\xi_1^{m_1}, \xi_2^{m_1}, \xi_3^{m_1})\\ \mathscr{D}^{\alpha}\xi_4^{m_2} &= \xi_5^{m_2}\\ \mathscr{D}^{\alpha}\xi_5^{m_2} &= \xi_6^{m_2}\\ \mathscr{D}^{\alpha}\xi_6^{m_2} &= -\mathscr{L}_{m_2}(\xi_4^{m_2}, \xi_5^{m_2}, \xi_6^{m_2})\end{aligned}$$

with

$$\mathscr{L}_{m_1}(\cdot) = -\frac{\rho\beta}{7}\xi_1^{m_1} - \left(\frac{8\rho - 7\beta}{7}\right)\xi_2^{m_1} - \left(\frac{\rho - 7}{7}\right)\xi_3^{m_1} + \frac{2\rho}{7}(\xi_1^{m_1} + \xi_2^{m_1} + \xi_3^{m_1})^3$$
$$\mathscr{L}_{m_2}(\cdot) = -(ac-1)\xi_5^{m_2} + c\xi_4^{m_2} + (c-a)\xi_6^{m_2} - (\xi_4^{m_2} - a\xi_5^{m_2} + \xi_6^{m_2})(\xi_5^{m_2} - a\xi_4^{m_2}) + b$$

and the family of slave systems in transformed coordinates is given by

$$\begin{aligned}
\mathscr{D}^\alpha \xi_1^{s_1} &= \xi_2^{s_1}\\
\mathscr{D}^\alpha \xi_2^{s_1} &= \xi_3^{s_1}\\
\mathscr{D}^\alpha \xi_3^{s_1} &= -\mathscr{L}_{s_1}(\xi_1^{s_1}, \xi_2^{s_1}, \xi_3^{s_1}, u_1^{s_1}, u_2^{s_1}, u_3^{s_1}) + \bar{u}_{s_1}\\
\mathscr{D}^\alpha u_1^{s_1} &= u_2^{s_1}\\
\mathscr{D}^\alpha u_2^{s_1} &= u_3^{s_1}\\
\mathscr{D}^\alpha u_3^{s_1} &= \bar{u}_{s_1}\\
\mathscr{D}^\alpha \xi_4^{s_2} &= \xi_5^{s_2}\\
\mathscr{D}^\alpha \xi_5^{s_2} &= \xi_6^{s_2}\\
\mathscr{D}^\alpha \xi_6^{s_2} &= -\mathscr{L}_{s_2}(\xi_4^{s_2}, \xi_5^{s_2}, \xi_6^{s_2}, u_4^{s_2}, u_5^{s_2}, u_6^{s_2}) + \bar{u}_{s_2}\\
\mathscr{D}^\alpha u_4^{s_2} &= u_5^{s_2}\\
\mathscr{D}^\alpha u_5^{s_2} &= u_6^{s_2}\\
\mathscr{D}^\alpha u_6^{s_2} &= \bar{u}_{s_2}\\
\mathscr{D}^\alpha \xi_7^{s_3} &= \xi_8^{s_3}\\
\mathscr{D}^\alpha \xi_8^{s_3} &= \xi_9^{s_3}\\
\mathscr{D}^\alpha \xi_9^{s_3} &= -\mathscr{L}_{s_3}(\xi_7^{s_3}, \xi_8^{s_3}, \xi_9^{s_3}, u_7^{s_3}, u_8^{s_3}, u_9^{s_3}) + \bar{u}_{s_3}\\
\mathscr{D}^\alpha u_7^{s_3} &= u_8^{s_3}\\
\mathscr{D}^\alpha u_8^{s_3} &= u_9^{s_3}\\
\mathscr{D}^\alpha u_9^{s_3} &= \bar{u}_{s_3}\\
\mathscr{D}^\alpha \xi_{10}^{s_4} &= \xi_{11}^{s_4}\\
\mathscr{D}^\alpha \xi_{11}^{s_4} &= \xi_{12}^{s_4}\\
\mathscr{D}^\alpha \xi_{12}^{s_4} &= -\mathscr{L}_{s_4}(\xi_{10}^{s_4}, \xi_{11}^{s_4}, \xi_{12}^{s_4}, u_{10}^{s_4}, u_{11}^{s_4}, u_{12}^{s_4}) + \bar{u}_{s_4}\\
\mathscr{D}^\alpha u_{10}^{s_4} &= u_{11}^{s_4}\\
\mathscr{D}^\alpha u_{11}^{s_4} &= u_{12}^{s_4}\\
\mathscr{D}^\alpha u_{12}^{s_4} &= \bar{u}_{s_4}\\
\mathscr{D}^\alpha \xi_{13}^{s_5} &= \xi_{14}^{s_5}\\
\mathscr{D}^\alpha \xi_{14}^{s_5} &= \xi_{15}^{s_5}
\end{aligned}$$

$$
\begin{aligned}
\mathscr{D}^\alpha \xi_{15}^{s_5} &= -\mathscr{L}_{s_5}(\xi_{13}^{s_5}, \xi_{14}^{s_5}, \xi_{15}^{s_5}, u_{13}^{s_5}, u_{14}^{s_5}, u_{15}^{s_5}) + \bar{u}_{s_5} \\
\mathscr{D}^\alpha u_{13}^{s_5} &= u_{14}^{s_5} \\
\mathscr{D}^\alpha u_{14}^{s_5} &= u_{15}^{s_5} \\
\mathscr{D}^\alpha u_{15}^{s_5} &= \bar{u}_{s_5}
\end{aligned}
$$

with

$$
\begin{aligned}
\mathscr{L}_{s_1}(\cdot) &= \beta_1(\xi_1^{s_1} - u_1^{s_1}) + \beta_2(\xi_2^{s_1} - u_2^{s_1}) + \beta_3(\xi_3^{s_1} - u_3^{s_1}) - \beta_4(\xi_1^{s_1} - u_1^{s_1})^3 \\
\mathscr{L}_{s_2}(\cdot) &= -\frac{\rho\beta}{7}(\xi_4^{s_2} - u_4^{s_2}) - \left(\frac{8\rho - 7\beta}{7}\right)(\xi_5^{s_2} - u_5^{s_2}) - \left(\frac{\rho - 7}{7}\right)(\xi_6^{s_2} - u_6^{s_2}) \\
&\quad + \frac{2\rho}{7}\left(\beta(\xi_4^{s_2} - u_4^{s_2}) + (\xi_5^{s_2} - u_5^{s_2}) + (\xi_6^{s_2} - u_6^{s_2})\right)^3 \\
\mathscr{L}_{s_3}(\cdot) &= \beta_1(\xi_7^{s_3} - u_7^{s_3}) + \beta_2(\xi_8^{s_3} - u_8^{s_3}) + \beta_3(\xi_9^{s_3} - u_9^{s_3}) - \beta_4(\xi_7^{s_3} - u_7^{s_3})^3 \\
\mathscr{L}_{s_4}(\cdot) &= -\frac{\rho\beta}{7}(\xi_{10}^{s_4} - u_{10}^{s_4}) - \left(\frac{8\rho - 7\beta}{7}\right)(\xi_{11}^{s_4} - u_{11}^{s_4}) - \left(\frac{\rho - 7}{7}\right)(\xi_{12}^{s_4} - u_{12}^{s_2}) \\
&\quad + \frac{2\rho}{7}\left(\beta(\xi_{10}^{s_4} - u_{10}^{s_4}) + (\xi_{11}^{s_4} - u_{11}^{s_4}) + (\xi_{12}^{s_4} - u_{12}^{s_4})\right)^3 \\
\mathscr{L}_{s_5}(\cdot) &= c(\xi_{13}^{s_5} - u_{13}^{s_5}) - (ac - 1)(\xi_{14}^{s_5} - u_{14}^{s_5}) + (c - a)(\xi_{15}^{s_5} - u_{15}^{s_5}) \\
&\quad - (\xi_{13}^{s_5} - u_{13}^{s_5} - a(\xi_{14}^{s_5} - u_{14}^{s_5}) + \xi_{15}^{s_5} - u_{15}^{s_5})(\xi_{14}^{s_5} - u_{14}^{s_5} - a(\xi_{13}^{s_5} - u_{13}^{s_5})) + b
\end{aligned}
$$

The closed-loop dynamics of synchronization error $e_\xi = \xi_m - \xi_s$ is given by the following augmented system

$$
\begin{aligned}
\mathscr{D}^\alpha e_{\xi_1}^{s_1} &= e_{\xi_2}^{s_1} \\
\mathscr{D}^\alpha e_{\xi_2}^{s_1} &= e_{\xi_3}^{s_1} \\
\mathscr{D}^\alpha e_{\xi_3}^{s_1} &= -\mathscr{L}_{m_1}(\xi_1^{m_1}, \xi_2^{m_1}, \xi_3^{m_1}) + \mathscr{L}_{s_1}(\xi_1^{s_1}, \xi_2^{s_1}, \xi_3^{s_1}, u_1^{s_1}, u_2^{s_1}, u_3^{s_1}) - \mathscr{D}^\alpha u_3^{s_1} \\
\mathscr{D}^\alpha u_1^{s_1} &= u_2^{s_1} \\
\mathscr{D}^\alpha u_2^{s_1} &= u_3^{s_1} \\
\mathscr{D}^\alpha u_3^{s_1} &= -\mathscr{L}_{m_1}(\xi_1^{m_1}, \xi_2^{m_1}, \xi_3^{m_1}) + \mathscr{L}_{s_1}(\xi_1^{s_1}, \xi_2^{s_1}, \xi_3^{s_1}, u_1^{s_1}, u_2^{s_1}, u_3^{s_1}) + k_{s_1} e_\xi^{s_1} \\
\mathscr{D}^\alpha e_{\xi_4}^{s_2} &= e_{\xi_5}^{s_2} \\
\mathscr{D}^\alpha e_{\xi_5}^{s_2} &= e_{\xi_6}^{s_2} \\
\mathscr{D}^\alpha e_{\xi_6}^{s_2} &= -\mathscr{L}_{m_1}(\xi_1^{m_1}, \xi_2^{m_1}, \xi_3^{m_1}) + \mathscr{L}_{s_2}(\xi_4^{s_2}, \xi_5^{s_2}, \xi_6^{s_2}, u_4^{s_2}, u_5^{s_2}, u_6^{s_2}) - \mathscr{D}^\alpha u_6^{s_2} \\
\mathscr{D}^\alpha u_4^{s_2} &= u_5^{s_2} \\
\mathscr{D}^\alpha u_5^{s_2} &= u_6^{s_2} \\
\mathscr{D}^\alpha u_6^{s_2} &= -\mathscr{L}_{m_1}(\xi_1^{m_1}, \xi_2^{m_1}, \xi_3^{m_1}) + \mathscr{L}_{s_2}(\xi_4^{s_2}, \xi_5^{s_2}, \xi_6^{s_2}, u_4^{s_2}, u_5^{s_2}, u_6^{s_2}) + k_{s_2} e_\xi^{s_2} \\
\mathscr{D}^\alpha e_{\xi_7}^{s_3} &= e_{\xi_8}^{s_3} \\
\mathscr{D}^\alpha e_{\xi_8}^{s_3} &= e_{\xi_9}^{s_3}
\end{aligned}
$$

$$
\begin{aligned}
\mathscr{D}^{\alpha} e^{s_3}_{\xi_9} &= -\mathscr{L}_{m_2}(\xi_4^{m_2}, \xi_5^{m_2}, \xi_6^{m_2}) + \mathscr{L}_{s_3}(\xi_7^{s_3}, \xi_8^{s_3}, \xi_9^{s_3}, u_7^{s_3}, u_8^{s_3}, u_9^{s_3}) - \mathscr{D}^{\alpha} u_9^{s_3} \\
\mathscr{D}^{\alpha} u_7^{s_3} &= u_8^{s_3} \\
\mathscr{D}^{\alpha} u_8^{s_3} &= u_9^{s_3} \\
\mathscr{D}^{\alpha} u_9^{s_3} &= -\mathscr{L}_{m_2}(\xi_4^{m_2}, \xi_5^{m_2}, \xi_6^{m_2}) + \mathscr{L}_{s_3}(\xi_7^{s_3}, \xi_8^{s_3}, \xi_9^{s_3}, u_7^{s_3}, u_8^{s_3}, u_9^{s_3}) + k_{s_3} e_{\xi}^{s_3} \\
\mathscr{D}^{\alpha} e^{s_4}_{\xi_{10}} &= e^{s_4}_{\xi_{11}} \\
\mathscr{D}^{\alpha} e^{s_4}_{\xi_{11}} &= e^{s_4}_{\xi_{12}} \\
\mathscr{D}^{\alpha} e^{s_4}_{\xi_{12}} &= -\mathscr{L}_{m_2}(\xi_4^{m_2}, \xi_5^{m_2}, \xi_6^{m_2}) + \mathscr{L}_{s_4}(\xi_{10}^{s_4}, \xi_{11}^{s_4}, \xi_{12}^{s_4}, u_{10}^{s_4}, u_{11}^{s_4}, u_{12}^{s_4}) - \mathscr{D}^{\alpha} u_{12}^{s_4} \\
\mathscr{D}^{\alpha} u_{10}^{s_4} &= u_{11}^{s_4} \\
\mathscr{D}^{\alpha} u_{11}^{s_4} &= u_{12}^{s_4} \\
\mathscr{D}^{\alpha} u_{12}^{s_4} &= -\mathscr{L}_{m_2}(\xi_4^{m_2}, \xi_5^{m_2}, \xi_6^{m_2}) + \mathscr{L}_{s_4}(\xi_{10}^{s_4}, \xi_{11}^{s_4}, \xi_{12}^{s_4}, u_{10}^{s_4}, u_{11}^{s_4}, u_{12}^{s_4}) + k_{s_4} e_{\xi}^{s_4} \\
\mathscr{D}^{\alpha} e^{s_5}_{\xi_{13}} &= e^{s_5}_{\xi_{14}} \\
\mathscr{D}^{\alpha} e^{s_5}_{\xi_{14}} &= e^{s_5}_{\xi_{15}} \\
\mathscr{D}^{\alpha} e^{s_5}_{\xi_{15}} &= -\mathscr{L}_{m_2}(\xi_4^{m_2}, \xi_5^{m_2}, \xi_6^{m_2}) + \mathscr{L}_{s_5}(\xi_{13}^{s_5}, \xi_{14}^{s_5}, \xi_{15}^{s_5}, u_{13}^{s_5}, u_{14}^{s_5}, u_{15}^{s_5}) - \mathscr{D}^{\alpha} u_{15}^{s_5} \\
\mathscr{D}^{\alpha} u_{13}^{s_5} &= u_{14}^{s_5} \\
\mathscr{D}^{\alpha} u_{14}^{s_5} &= u_{15}^{s_5} \\
\mathscr{D}^{\alpha} u_{15}^{s_5} &= -\mathscr{L}_{m_2}(\xi_4^{m_2}, \xi_5^{m_2}, \xi_6^{m_2}) + \mathscr{L}_{s_5}(\xi_{13}^{s_5}, \xi_{14}^{s_6}, \xi_{16}^{s_6}, u_{13}^{s_5}, u_{14}^{s_5}, u_{15}^{s_5}) + k_{s_5} e_{\xi}^{s_5}
\end{aligned}
$$

After some algebraic manipulations, it is obtained $\mathscr{D}^{\alpha} e_{\xi} = (\mathscr{A} - \mathscr{K}) e_{\xi}$. Then, synchronization error converges asymptotically to zero if matrix

$$
\mathscr{A} - \mathscr{K} = diag(\bar{\mathscr{A}_1}, \bar{\mathscr{A}_2}, \bar{\mathscr{A}_3}, \bar{\mathscr{A}_4}, \bar{\mathscr{A}_5}) \tag{7.22}
$$

is Hurwitz. Where

$$
\bar{\mathscr{A}_j} = \begin{bmatrix} 0 & 1 & 0 \\ 0 & 0 & 1 \\ -k_{1,j} & -k_{2,j} & -k_{3,j} \end{bmatrix}, \quad 1 \le j \le 5.
$$

Parameters are taken as in the first example, with $a = 0.5$, $b = 0.2$ and $c = 10$ [25]. Let initial conditions be $x_{m_1}(0) = \left[-0.50\ -0.07\ 0.65\right]^T$, $x_{m_2}(0) = \left[0.50\ 1.5\ 0.1\right]^T$, $x_{s_1}(0) = \left[-0.20\ 0.35\ 0.20\right]^T$, $x_{s_2}(0) = \left[-0.58\ -0.01\ 0.30\right]^T$, $x_{s_3}(0) = \left[2\ -0.1\ -2\right]^T$, $x_{s_4}(0) = \left[-0.71\ -0.1\ 0.45\right]^T$, $x_{s_5}(0) = \left[1\ 2.5\ -1\right]^T$, to ensure chaotic behavior and $k_{s,j} = \left[10\ 20\ 10\right]$ for $1 \le j \le 5$.

Figure 7.9 illustrates synchronization error convergence to the origin in transformed and original coordinates, respectively; FGMS is shown in Figs. 7.10 and 7.11 in transformed coordinates. And taking the mapping (7.23) FGMS is shown in original coordinates in Figs. 7.12 and 7.13. The effectiveness of our approach can be verified from the multi-synchronization trajectories. Here, multi-synchronization

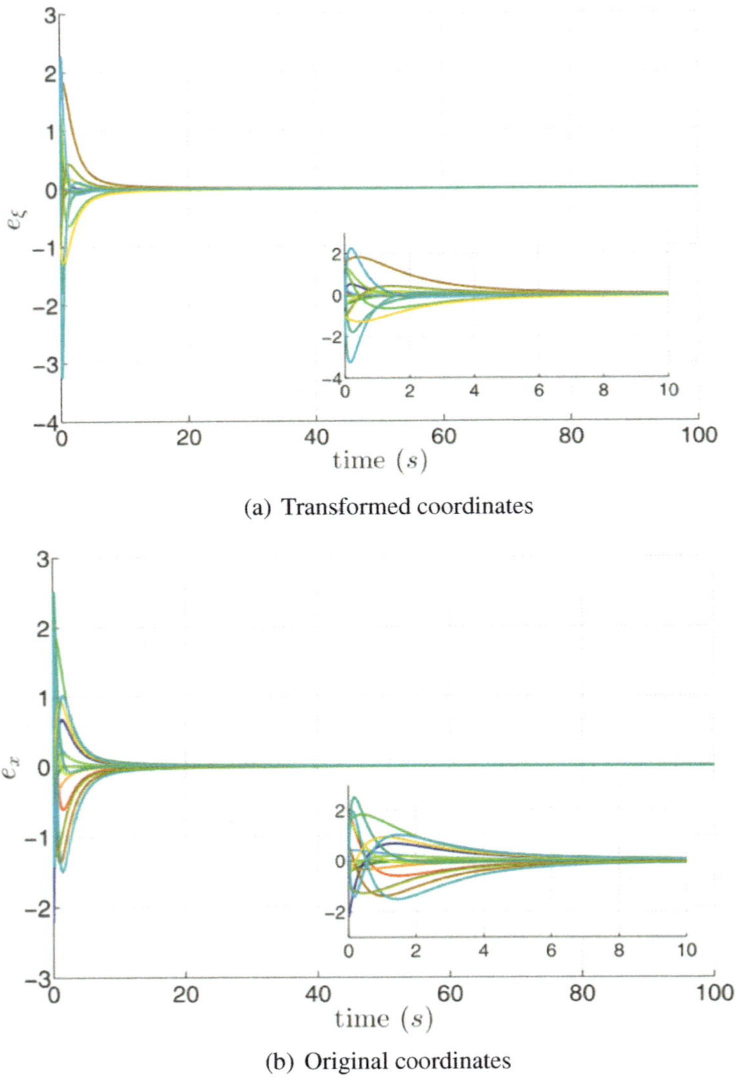

Fig. 7.9 Time evolution of synchronization error (family of systems)

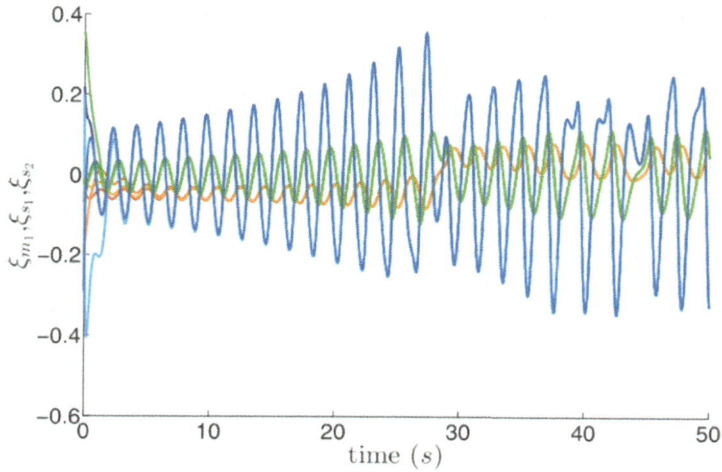

(a) Time evolution of synchronized trajectories

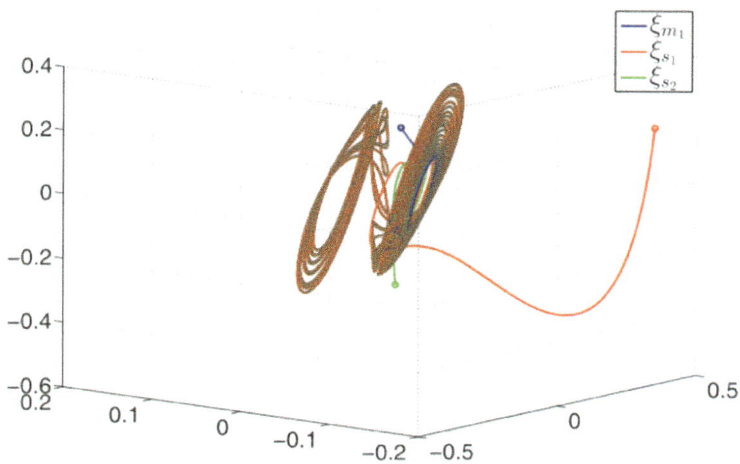

(b) Multi-synchronization trajectories

Fig. 7.10 FGMS in transformed coordinates (family of systems, 1st group)

is shown for two decoupled groups. Note that the synchronization error converges asymptotically to the origin. In the case of the first group, convergence to the trajectories of first master system is more slowly than Example 7.2 (see Figs. 7.6, 7.7, 7.10 and 7.12, respectively). This is due gains $k_{s,1}$, $k_{s,2}$ that are smaller than gains in Example 7.2. It is worth mentioning that our methodology can be applied not only to Liouvillian systems, and this is the case of any system that fulfills FAO condition, i.e., Rössler system.

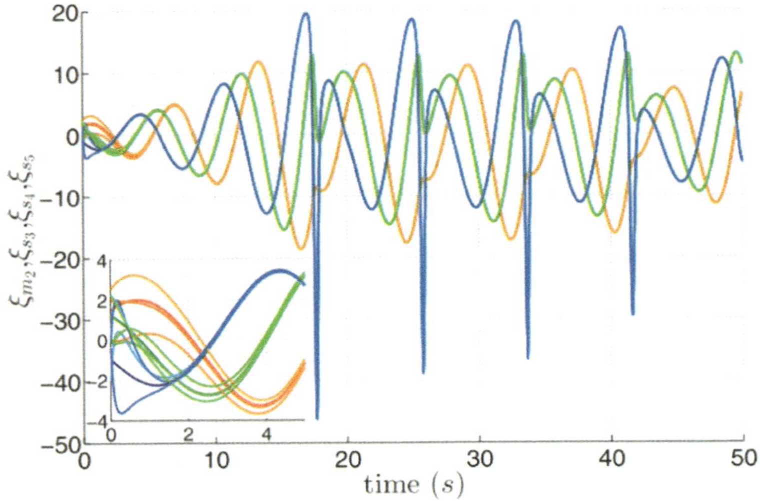

(a) Time evolution of synchronized trajectories

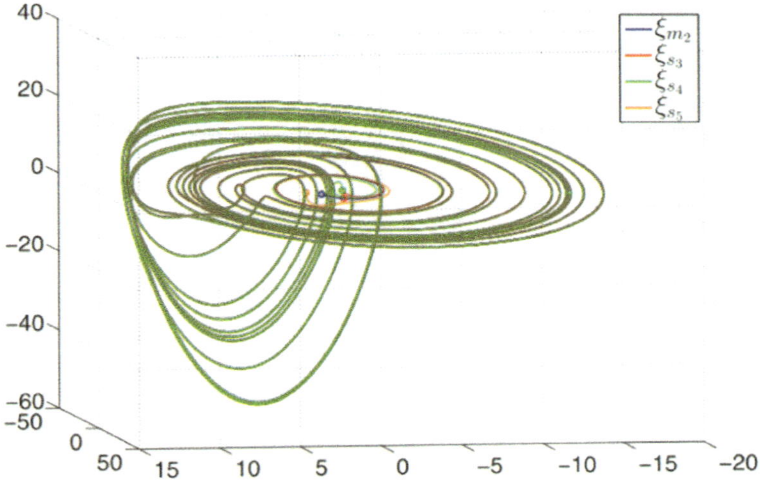

(b) Multi-synchronization trajectories

Fig. 7.11 FGMS in transformed coordinates (family of systems, 2nd group)

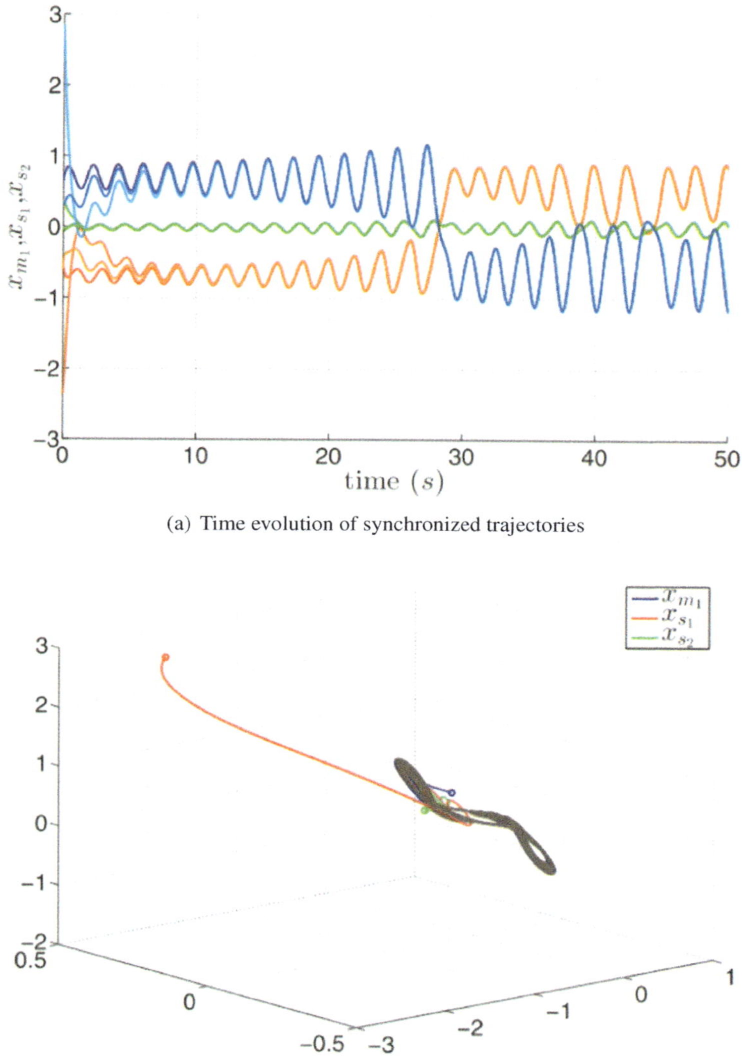

(a) Time evolution of synchronized trajectories

(b) Multi-synchronization trajectories

Fig. 7.12 FGMS in original coordinates (family of systems, 1st group)

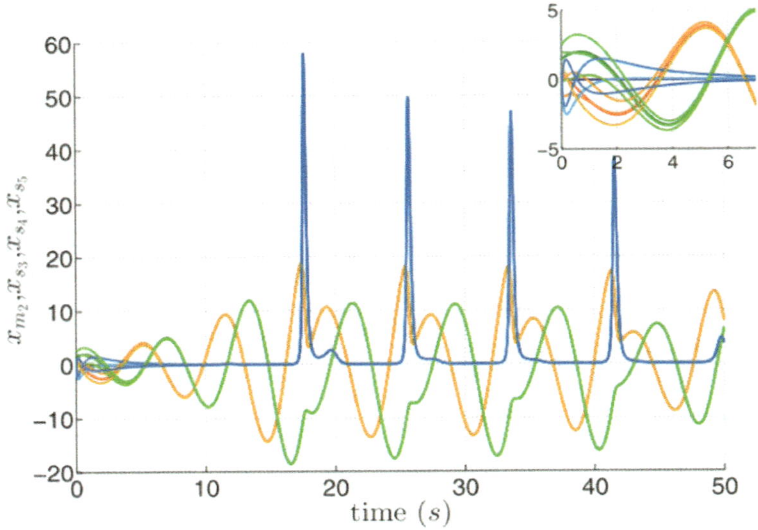

(a) Time evolution of synchronized trajectories

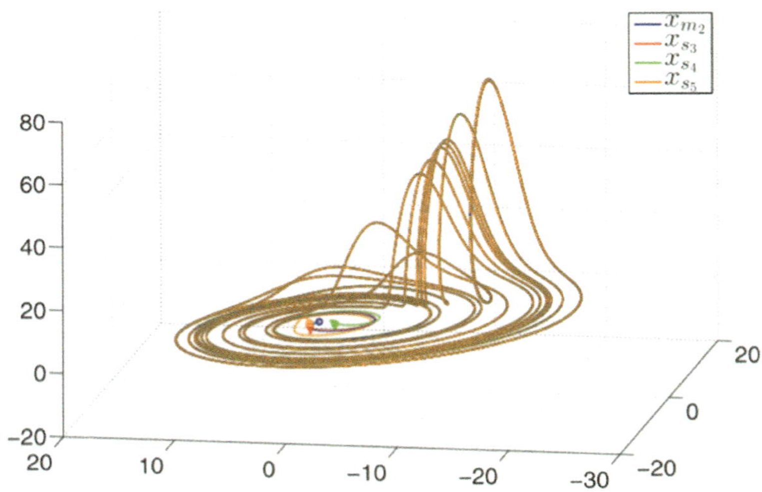

(b) Multi-synchronization trajectories

Fig. 7.13 FGMS in original coordinates (family of systems, 2nd group)

$$H_{ms}(X_s) = \begin{pmatrix} \beta(x_1^{s_1} + u_1^{s_1}) + x_2^{s_1} + u_2^{s_1} + x_3^{s_1} + u_3^{s_1} \\ x_2^{s_1} + u_2^{s_1} \\ -\beta(x_1^{s_1} + u_1^{s_1}) \\ x_4^{s_2} + \beta u_4^{s_2} + u_5^{s_2} + u_6^{s_2} \\ x_5^{s_2} + u_5^{s_2} \\ -\beta(x_4^{s_2} - x_5^{s_2} + x_6^{s_2} + u_6^{s_2}) \\ x_8^{s_3} - a x_7^{s_3} \\ x_7^{s_3} \\ -x_7^{s_3} + a x_8^{s_3} - x_9^{s_3} \\ x_{11}^{s_4} + \frac{a}{\beta} x_{12}^{s_4} \\ -\frac{1}{\beta} x_{12}^{s_4} \\ -x_{10}^{s_4} + (a+1) x_{11}^{s_4} + \left(\frac{1}{\beta} - 1\right) x_{12}^{s_4} \\ x_{13}^{s_5} \\ x_{14}^{s_5} \\ x_{15}^{s_5} \end{pmatrix} \tag{7.23}$$

7.3 Generalized Synchronization of PDE Systems of Fractional Order

The models based on partial differential equations are extremely helpful and accurate for modeling real-world phenomena. Nonetheless, as in the integer-order case, there exists a wide variety of complex phenomena, where differential equations of non integer order fit better the experimental data or allow to obtain models with less parameters. For example, the so-called anomalous diffusion process is better describe by fractional-order PDE systems. Moreover, it is well known that the visco-elastic behavior of materials can be modeled by complex spring and dashpot arrangements. However, due to its complexity, unlike fractional-order models, these require to identify a considerable number of parameters, which is not practical in most cases.

Therefore, in the following the problem of GS for fractional-order PDE systems is addressed. First of all, let us consider the following properties and definitions.

Definition 7.8. ([30]) A real function $f(t)$, $t > 0$, is said to be in the space C_μ, $\mu \in \mathbb{R}$ if there exists a real number $\rho > \mu$, such that $f(t) = t^\rho f_1(t)$, where $f_1(t) \in C[0, \infty)$, and it is said to be in the space C_μ^η iff $f^{(\eta)} \in C_\mu$, $\eta \in \mathbb{N}$.

For the Riemann-Liouville fractional integral, the following are some relevant properties.

For $f \in C_\mu$, $\mu \geq -1$, $\alpha, \beta \geq 0$ and $\gamma > -1$:

(1a) $I^\alpha I^\beta f(t) = I^{\alpha+\beta} f(t)$,
(2a) $I^\alpha I^\beta f(t) = I^\beta I^\alpha f(t)$,
(3a) $I^\alpha t^\gamma = \frac{\Gamma(\gamma+1)}{\Gamma(\alpha+\gamma+1)} t^{\alpha+\gamma}$.

These properties and some others can be found in [26]. On the other hand, the following are some properties of interest of the Caputo fractional derivative.

If $\eta - 1 < \alpha \leq \eta$, $\eta \in \mathbb{N}$ and $f \in C^{\eta}_{\mu}$, $\mu \geq -1$, then

(1b) $D^{\alpha}_{C} I^{\alpha} f(t) = f(t)$,
(2b) $I^{\alpha} D^{\alpha}_{C} f(t) = f(t) - \sum_{k=0}^{\eta-1} f^{(k)}(0^{+}) \frac{t^k}{k!}$, $t > 0$,
(3b) $D^{\alpha}_{C} (f(t) + g(t)) = D^{\alpha}_{C} f(t) + D^{\alpha}_{C} g(t)$,
(4b) $D^{\alpha}_{C} (f(t)g(t)) = \sum_{k=0}^{\infty} \binom{\alpha}{k} \left(D^{\alpha-k} f(t)\right) g^{(k)}(t) - \sum_{k=0}^{\eta-1} \frac{t^{k-\alpha}}{\Gamma(k+1-\alpha)} \big[(f(t)g(t))^{(k)} (0) \big]$,
(5b) $D^{\alpha}_{C} a = 0$ for $a = const.$

Definition 7.9. ([21] *Let* $v(x, t)) : \mathbb{R} \times [0, \infty) \to \mathbb{R}$ be a causal function of time, i.e., $v(x, t) = 0$ for $t < 0$, and η the smallest integer that exceeds α. The Caputo time-fractional derivative of order α of $v(x, t)$ is defined as

$$
\begin{aligned}
D^{\alpha}_{Ct} v(x, t) &= \frac{\partial^{\alpha} v(x, t)}{\partial t^{\alpha}} \\
&= \begin{cases} \frac{1}{\Gamma(\eta-\alpha)} \int_0^t (t - \tau)^{\eta-\alpha-1} \frac{\partial^{\eta} v(x,\tau)}{\partial \tau^{\eta}} d\tau, \\ for \quad \eta - 1 < \alpha < \eta, \\ \\ \frac{\partial^{\eta} v(x,t)}{\partial t^{\eta}}, \quad for \quad \alpha = \eta \in \mathbb{N}. \end{cases}
\end{aligned}
\tag{7.24}
$$

7.3.1 Fractional-Order Dynamical Distributed Controller

The synchronization of two fractional-order PDE-based systems can be reduced to a problem of *fractional generalized synchronization* (FGS), where is sufficient to know the output of each system. To solve the FGS problem, a dynamical distributed controller of fractional order can be implemented. Such controller is obtained from canonical forms of the PDE-based systems.

Let us recall the sequential operator (7.3), i.e.,

$$
\mathscr{D}^{(r\alpha)} v(x, t) = \underbrace{D^{\alpha}_{Ct} D^{\alpha}_{Ct} \ldots D^{\alpha}_{Ct}}_{r\text{-times}} v(x, t), \quad r \in \mathbb{N}, \tag{7.25}
$$

We can note that if $r = 1$ then $\mathscr{D}^{(\alpha)} v(x, t) = D^{\alpha}_{Ct} v(x, t) = \frac{\partial^{\alpha} v(x,t)}{\partial t^{\alpha}}$.

By considering the last operator, a fractional-order PDE system can be expressed as:

$$
\begin{aligned}
\mathscr{D}^{(\alpha)} V(x, t) &= R[x] V(x, t) + Q(x, t, u), \quad x \in [0, L], \\
y(x, t) &= h(V, u),
\end{aligned}
\tag{7.26}
$$

where $V(x,t) = [v_1(x,t), v_2(x,t), \ldots, v_n(x,t)]^T$ is the state vector, $y(x,t) \in \mathbb{R}^p$ the output and $u(x,t)$ the input. The operator $R[x]$ is a differential operator of integer order in x and of proper dimensions. The functions $h(\cdot)$ and $Q(\cdot)$ are continuous.

The analytic solution of system (7.26) can be approximated by means of the Variational Iteration Method (VIM). Thus, the approximate solution is given by:

$$V_{k+1}(x,t) = V_k(x,t) - I_t^{\alpha}\left[\mathscr{D}^{(\alpha)}V_k(x,t) - R[x]V_k(x,t) - Q(x,t,u)\right] \quad (7.27)$$

where I_t^{α} is the Riemann-Liouville fractional integral with respect to t. The initial condition $V_0(x,t)$ can be used as a first approximation (for further details see [6, 12]).

Let $n \geq 0$ be the minimum integer such that $\mathscr{D}^{(n\alpha)}y$ is analytically dependent on $\{y, \mathscr{D}^{(\alpha)}y, \mathscr{D}^{(2\alpha)}y, \ldots, \mathscr{D}^{([n-1]\alpha)}y\}$, such that we have the following input-output system:

$$H\left(y, \mathscr{D}^{(\alpha)}y, \ldots, \mathscr{D}^{([n-1]\alpha)}y, \mathscr{D}^{(n\alpha)}y, u, \mathscr{D}^{(\alpha)}u, \ldots, \mathscr{D}^{([\varphi-1]\alpha)}u, \mathscr{D}^{(\varphi\alpha)}u\right) = 0, \quad (7.28)$$

the system (7.28) can be locally solved as

$$\mathscr{D}^{(n\alpha)}y = \mathscr{L}\left(y, \mathscr{D}^{(\alpha)}y, \ldots, \mathscr{D}^{([n-1]\alpha)}y, u, \mathscr{D}^{(\alpha)}u, \ldots, \mathscr{D}^{([\varphi-1]\alpha)}u\right) + \mathscr{D}^{(\varphi\alpha)}u, \quad (7.29)$$

where the integer $\varphi \geq 0$. Now, consider the change of variable $z_i = \mathscr{D}^{([i-1]\alpha)}y$, $1 \leq i \leq n$, such that we can obtain the following canonical form:

$$\begin{aligned}
\mathscr{D}^{(\alpha)}z_1 &= z_2 \\
\mathscr{D}^{(\alpha)}z_2 &= z_3 \\
&\vdots \\
\mathscr{D}^{(\alpha)}z_{n-1} &= z_n \\
\mathscr{D}^{(\alpha)}z_n &= \mathscr{L}\left(z_1, z_2, \ldots, z_n, u, \mathscr{D}^{(\alpha)}u, \ldots, \mathscr{D}^{([\varphi-1]\alpha)}u\right) + \mathscr{D}^{(\varphi\alpha)}u \\
\bar{y} &= z_1.
\end{aligned} \quad (7.30)$$

where $\bar{y}$ is the output and $Z = [z_1, z_2, \ldots, z_n]^T$ is the new state vector. The representation above is the fractional generalized observability canonical form (FGOCF).

Let us consider two fractional-order PDE-based systems, master and slave. The master fractional PDE-based system is given by:

$$\begin{aligned}
\mathscr{D}^{(\alpha)}V(x,t) &= R_m[x]V(x,t) + Q_m(x,t,u_m), \quad x \in [0, L], \\
y_m(x,t) &= h_m(V, u_m),
\end{aligned} \quad (7.31)$$

where $V(x,t) \in \mathbb{R}^{n_m}$. On the other hand, the slave system is:

$$\mathscr{D}^{(\alpha)} W(x,t) = R_s[x] W(x,t) + Q_s(x,t,u_s), \quad x \in [0,L],$$
$$y_s(x,t) = h_s(W,u_s), \tag{7.32}$$

where $W(x,t) \in \mathbb{R}^{n_s}$. Here, sub-indexes m and s refer to "master" and "slave", respectively.

Definition 7.10. The master system (7.31) and slave system (7.32) are said to be in a state of fractional generalized synchronization (FGS) if there exists a change of variable that generates a transformation $H_{ms} : \mathbb{R}^{n_s} \to \mathbb{R}^{n_m}$ with $H_{ms} = \Phi_s^{-1} \circ \Phi_m$, an algebraic manifold $\mathscr{M} = \{(W,V) | V = H_{ms}(W)\}$ and a compact set $B \subset \mathbb{R}^{n_s} \times \mathbb{R}^{n_m}$, with $\mathscr{M} \subset B$, such that all state trajectory with initial conditions in B approach $\mathscr{M}$ as $t \to \infty$, i.e., if

$$\lim_{t \to \infty} \| H_{ms}(W) - V \| = 0. \tag{7.33}$$

The transformations $\Phi_m : \mathbb{R}^{n_m} \to \mathbb{R}^n$ and $\Phi_s : \mathbb{R}^{n_s} \to \mathbb{R}^n$ have a common state space, i.e., $\Phi_m(V) = Z_m \in \mathbb{R}^n$ and $\Phi_s(W) = Z_s \in \mathbb{R}^n$, and can be found by using a change of variable that we can define as a linear combination of the available variables, i.e.:

$$z = \sum_i a_i v_i + \sum_j b_j u_j, \quad a_i, b_j \in \mathbb{R}. \tag{7.34}$$

Theorem 7.3. *Let the systems* (7.31) *and* (7.32) *be transformable to a FGOCF. Then, there exists a fractional dynamical distributed controller such that*

$$\lim_{t \to \infty} \| Z_m - Z_s \| = 0, \tag{7.35}$$

where $Z_m = \left[z_{m_1}, z_{m_2}, \ldots, z_{m_n}\right]^T$ *and* $Z_s = \left[z_{s_1}, z_{s_2}, \ldots, z_{s_n}\right]^T$ *are the state trajectories of master and slave systems in the new coordinates, respectively. With* $z_{m_i} = \mathscr{D}^{([i-1]\alpha)} y_m$ *and* $z_{s_i} = \mathscr{D}^{([i-1]\alpha)}(y_s + u_s)$*, for* $1 \le i \le n$.

Proof. Let us choose the following change of variable

$$z_{m_1} = y_m, \tag{7.36}$$

and

$$z_{s_1} = y_s + u_s. \tag{7.37}$$

for master and slave system, respectively. Then, we can generate the coordinate transformations

$$\Phi_m = \begin{bmatrix} z_{m_1} \\ \mathscr{D}^{(\alpha)} z_{m_1} \\ \mathscr{D}^{(2\alpha)} z_{m_1} \\ \vdots \\ \mathscr{D}^{([n-1]\alpha)} z_{m_1} \end{bmatrix}, \qquad \Phi_s = \begin{bmatrix} z_{s_1} \\ \mathscr{D}^{(\alpha)} z_{s_1} \\ \mathscr{D}^{(2\alpha)} z_{s_1} \\ \vdots \\ \mathscr{D}^{([n-1]\alpha)} z_{s_1} \end{bmatrix},$$

Thus, the FGOCF of the master system (7.31) is:

$$\begin{aligned} \mathscr{D}^{(\alpha)} z_{m_j} &= z_{m_{j+1}}, \quad 1 \le j \le n-1, \\ \mathscr{D}^{(\alpha)} z_{m_n} &= \mathscr{L}_m \left(z_{m_1}, z_{m_2}, \ldots, z_{m_n} \right), \end{aligned} \tag{7.38}$$

and of the slave system (7.32) is:

$$\begin{aligned} \mathscr{D}^{(\alpha)} z_{s_j} &= z_{s_{j+1}}, \quad 1 \le j \le n-1, \\ \mathscr{D}^{(\alpha)} z_{s_n} &= \mathscr{L}_s \left(z_{s_1}, z_{s_2}, \ldots, z_{s_n}, u_s, \mathscr{D}^{(\alpha)} u_s, \ldots, \mathscr{D}^{([\gamma-1]\alpha)} u_s \right) + \mathscr{D}^{(\gamma\alpha)} u_s. \end{aligned} \tag{7.39}$$

Let $e = Z_m - Z_s$ be the synchronization error, and consider the following notation $u_1 = u_s, u_2 = \mathscr{D}^{(\alpha)} u_s, \ldots, u_\gamma = \mathscr{D}^{([\gamma-1]\alpha)} u_s$. Then, from (7.38) and (7.39) we have

$$\begin{aligned} \mathscr{D}^{(\alpha)} e_j &= e_{j+1}, \quad 1 \le j \le n-1, \\ \mathscr{D}^{(\alpha)} e_n &= \mathscr{L}_m \left(z_{m_1}, z_{m_2}, \ldots, z_{m_n} \right) - \mathscr{L}_s \left(z_{s_1}, z_{s_2}, \ldots, z_{s_n}, u_1, u_2, \ldots, u_\gamma \right) \\ &\quad - \mathscr{D}^{(\alpha)} u_\gamma. \end{aligned} \tag{7.40}$$

In such a way that we propose the following fractional dynamical distributed controller in order to synchronize the fractional PDE-based systems:

$$\begin{aligned} \mathscr{D}^{(\alpha)} u_j &= u_{j+1}, \quad 1 \le j \le \gamma - 1, \\ \mathscr{D}^{(\alpha)} u_\gamma &= \mathscr{L}_m \left(z_{m_1}, z_{m_2}, \ldots, z_{m_n} \right) - \mathscr{L}_s \left(z_{s_1}, z_{s_2}, \ldots, z_{s_n}, u_1, u_2, \ldots, u_\gamma \right) \\ &\quad + \kappa e. \end{aligned} \tag{7.41}$$

where $\kappa = [\kappa_1, \kappa_2, \ldots, \kappa_n]^T$. Therefore, the dynamics of the synchronization error is given by the following augmented system:

$$\begin{aligned} \mathscr{D}^{(\alpha)} e_j &= e_{j+1}, \quad 1 \le j \le n-1, \\ \mathscr{D}^{(\alpha)} e_n &= \mathscr{L}_m \left(z_{m_1}, z_{m_2}, \ldots, z_{m_n} \right) - \mathscr{L}_s \left(z_{s_1}, z_{s_2}, \ldots, z_{s_n}, u_1, u_2, \ldots, u_\gamma \right) \\ &\quad - \mathscr{D}^{(\alpha)} u_\gamma, \\ \mathscr{D}^{(\alpha)} u_j &= u_{j+1}, \quad 1 \le j \le \gamma - 1, \\ \mathscr{D}^{(\alpha)} u_\gamma &= \mathscr{L}_m \left(z_{m_1}, z_{m_2}, \ldots, z_{m_n} \right) - \mathscr{L}_s \left(z_{s_1}, z_{s_2}, \ldots, z_{s_n}, u_1, u_2, \ldots, u_\gamma \right) \\ &\quad + \kappa e. \end{aligned} \tag{7.42}$$

Then, the closed-loop system is given by the following fractional-order PDE-based system:

$$\mathscr{D}^{(\alpha)} e = Ae, \tag{7.43}$$

where

$$A = \begin{bmatrix} 0 & 1 & 0 & \dots & 0 & 0 \\ 0 & 0 & 1 & \dots & 0 & 0 \\ \vdots & \vdots & \vdots & \ddots & \vdots & \vdots \\ 0 & 0 & 0 & \dots & 1 & 0 \\ 0 & 0 & 0 & \dots & 0 & 1 \\ -\kappa_1 & -\kappa_2 & -\kappa_3 & \dots & -\kappa_{n-1} & -\kappa_n \end{bmatrix} \tag{7.44}$$

We can determine the stability of the above system by considering the result obtained by Matignon [20], i.e., the asymptotic stability of the fractional-order PDE-based system (7.43) is guaranteed if

$$|arg\ (spec(A))\ | > \frac{\alpha\pi}{2}$$

We can notice that A is a linear bounded operator, whose spectrum $spec(A)$ is the finite set $\bar{\lambda} = \{\lambda_i \quad | \quad 1 \leq i \leq n\}$, where λ_i is an eigenvalue of matrix A. Therefore, system (7.43) will be asymptotically stable if the controller gains κ_i are such that for every λ_i we have

$$|arg\ (\lambda_i)\ | > \frac{\alpha\pi}{2}.$$

□

7.3.2 FGS of Schnakenberg Systems

The Schnakenberg system is a simple reaction-diffusion system often described as a substrate-depletion model. It was introduced in 1979 by J. Schnakenberg to show sustained oscillations in a simple model of glycolysis (conversion of glucose to energy) [31]. The fractional version of this system is given by:

$$\begin{aligned} \mathscr{D}^{(\alpha)} v_1(x,t) &= \frac{\partial^2 v_1(x,t)}{\partial x^2} + a_m - v_1(x,t) + v_1^2(x,t) v_2(x,t), \\ \mathscr{D}^{(\alpha)} v_2(x,t) &= d_m \frac{\partial^2 v_2(x,t)}{\partial x^2} + b_m - v_1^2(x,t) v_2(x,t), \\ y_m &= v_1(x,t). \end{aligned} \tag{7.45}$$

Let system (7.45) be the master system. On the other hand, let us consider a second Schnakenberg system (slave system) denoted as:

$$\begin{aligned}
\mathscr{D}^{(\alpha)}\omega_1(x,t) &= \frac{\partial^2\omega_1(x,t)}{\partial x^2} + a_s - \omega_1(x,t) + \omega_1^2(x,t)\omega_2(x,t),\\
\mathscr{D}^{(\alpha)}\omega_2(x,t) &= d_s\frac{\partial^2\omega_2(x,t)}{\partial x^2} + b_s - \omega_1^2(x,t)\omega_2(x,t),\\
y_s &= \omega_1(x,t).
\end{aligned} \tag{7.46}$$

where ν_1, ω_1 are activator concentrations and ν_2, ω_2 are substrate concentrations of master and slave system, respectively. The constants a_m, a_s and b_m, b_s are speed rate parameters, and $d_m = d_{m_1}/d_{m_2}$, $d_s = d_{s_1}/d_{s_2}$ are diffusion coefficient ratios.

Then, let $z_{m_1} = y_m$ and $z_{s_1} = y_s + u_1$ be a change of variable for master and slave system, respectively. Thus, we can generate the transformations:

$$\Phi_m = \begin{pmatrix} z_{m_1} \\ \mathscr{D}^{(\alpha)}z_{m_1} \end{pmatrix} = \begin{pmatrix} y_m \\ \mathscr{D}^{(\alpha)}y_m \end{pmatrix} \tag{7.47}$$

and

$$\Phi_s = \begin{pmatrix} z_{s_1} \\ \mathscr{D}^{(\alpha)}z_{s_1} \end{pmatrix} = \begin{pmatrix} y_s + u_1 \\ \mathscr{D}^{(\alpha)}y_s + u_2 \end{pmatrix}. \tag{7.48}$$

Such that, master system (7.45) is expressed as:

$$\begin{aligned}
\mathscr{D}^{(\alpha)}z_{m_1} &= z_{m_2}\\
\mathscr{D}^{(\alpha)}z_{m_2} &= L_{m_1} + L_{m_2} + L_{m_3} = \mathscr{L}_m\left(z_{m_1}, z_{m_2}\right),
\end{aligned} \tag{7.49}$$

where

$$\begin{aligned}
L_{m_1} &= \frac{1}{\Gamma(1-\alpha)}\int_0^t (t-\tau)^{-\alpha}\frac{\partial^3\nu_1(x,t)}{\partial x^2\partial\tau}d\tau,\\
L_{m_2} &= \nu_1(x,t) - \nu_1^2(x,t)\nu_2(x,t) - \frac{\partial^2\nu_1(x,t)}{\partial x^2} - a_m,\\
L_{m_3} &= \sum_{k=0}^{\infty}\binom{\alpha}{k}(D^{\alpha-k}\nu_1^2)\nu_2^{(k)} - \frac{t^{-1/2}}{\Gamma(1/2)}(\nu_1^2\nu_2)(x,0).
\end{aligned}$$

On the other hand, the FGOCF of slave system (7.46) is:

$$\begin{aligned}
\mathscr{D}^{(\alpha)}z_{s_1} &= z_{s_2}\\
\mathscr{D}^{(\alpha)}z_{s_2} &= L_{s_1} + L_{s_2} + L_{s_3} + \mathscr{D}^{(\alpha)}u_2 = \mathscr{L}_s\left(z_{s_1}, z_{s_2}, u_1, u_2\right) + \mathscr{D}^{(\alpha)}u_2,
\end{aligned} \tag{7.50}$$

where

$$L_{s_1} = \frac{1}{\Gamma(1-\alpha)}\int_0^t (t-\tau)^{-\alpha}\frac{\partial^3\omega_1(x,t)}{\partial x^2\partial\tau}d\tau,$$
$$L_{s_2} = \omega_1(x,t) - \omega_1^2(x,t)\omega_2(x,t) - \frac{\partial^2\omega_1(x,t)}{\partial x^2} - a_s,$$
$$L_{s_3} = \sum_{k=0}^{\infty}\binom{\alpha}{k}(D^{\alpha-k}\omega_1^2)\omega_2^{(k)} - \frac{t^{-1/2}}{\Gamma(1/2)}(\omega_1^2\omega_2)(x,0).$$

Hence, we have:

$$\begin{aligned}
\mathscr{D}^{(\alpha)}e_1 &= e_2 \\
\mathscr{D}^{(\alpha)}e_2 &= \mathscr{L}_m\left(z_{m_1}, z_{m_2}\right) - \mathscr{L}_s\left(z_{s_1}, z_{s_2}, u_1, u_2\right) - \mathscr{D}^{(\alpha)}u_2 \\
\mathscr{D}^{(\alpha)}u_1 &= u_2 \\
\mathscr{D}^{(\alpha)}u_2 &= \mathscr{L}_m\left(z_{m_1}, z_{m_2}\right) - \mathscr{L}_s\left(z_{s_1}, z_{s_2}, u_1, u_2\right) + \kappa e,
\end{aligned} \tag{7.51}$$

with $\kappa = [\kappa_1, \kappa_2]^T$. Then, the closed-loop system is:

$$\mathscr{D}^{(\alpha)}e = Ae, \tag{7.52}$$

where

$$A = \begin{bmatrix} 0 & 1 \\ -\kappa_1 & -\kappa_2 \end{bmatrix} \tag{7.53}$$

Therefore, the system is asymptotically stable if the controller gains κ_1 and κ_2 satisfy the following

$$|arg\left(-\frac{1}{2}\left(\kappa_2 \pm \sqrt{\kappa_2^2 - 4\kappa_1}\right)\right)| > \frac{\alpha\pi}{2}$$

For the numerical simulation, we consider $\kappa_1 = 70$ and $\kappa_2 = 50$, along with the parameters $a_m = 0.1$, $b_m = 0.9$, $d_m = 0.25$, $a_s = 0.3$, $b_s = 0.5$ and $d_s = 0.15$. Besides, the initial conditions are as follows:

$$\nu_1(x,0) = \sin(x), \quad \nu_2(x,0) = \cos(x).$$
$$\omega_1(x,0) = e^{1/x}, \quad \omega_2(x,0) = \sin(x) + \cos(x).$$

By means of the Variational Iteration Method the approximate solutions are

$$\begin{aligned}
\nu_{1_{k+1}}(x,t) = \nu_{1_k}(x,t) - J_t^{\alpha}\Big[\mathscr{D}^{(\alpha)}\nu_{1_k}(x,t) - \frac{\partial^2 \nu_{1_k}(x,t)}{\partial x^2} \\
-a_m + \nu_{1_k}(x,t) - \nu_{1_k}^2(x,t)\nu_{2_k}(x,t)\Big], \\
\nu_{2_{k+1}}(x,t) = \nu_{2_k}(x,t) - J_t^{\alpha}\Big[\mathscr{D}^{(\alpha)}\nu_{2_k}(x,t) - d_m\frac{\partial^2 \nu_{2_k}(x,t)}{\partial x^2} \\
-b_m + \nu_{1_k}^2(x,t)\nu_{2_k}(x,t)\Big],
\end{aligned}$$

where $\nu_{1_0}(x,t) = \sin(x)$ and $\nu_{2_0}(x,t) = \cos(x)$. And

$$\begin{aligned}
\omega_{1_{k+1}}(x,t) =& \omega_{1_k}(x,t) - J_t^{\alpha}\Big[\mathscr{D}^{(\alpha)}\omega_{1_k}(x,t) - \frac{\partial^2 \omega_{1_k}(x,t)}{\partial x^2} \\
&-a_s + \omega_{1_k}(x,t) - \omega_{1_k}^2(x,t)\omega_{2_k}(x,t)\Big], \\
\omega_{2_{k+1}}(x,t) =& \omega_{2_k}(x,t) - J_t^{\alpha}\Big[\mathscr{D}^{(\alpha)}\omega_{2_k}(x,t) - d_s\frac{\partial^2 \omega_{2_k}(x,t)}{\partial x^2} \\
&-b_s + \omega_{1_k}^2(x,t)\omega_{2_k}(x,t)\Big],
\end{aligned}$$

where we use the corresponding initial conditions as first approximation, i.e., $\omega_{1_0}(x,t) = e^{1/x}$ and $\omega_{2_0}(x,t) = \sin(x) + \cos(x)$. These iterations can be easily computed with a symbolic calculation software. We consider the fractional order $\alpha = 1/2$.

In Figs. 7.14 and 7.15 we can observe the behavior in the original coordinates of the Schnakenberg master and slave system, respectively. We can notice that these are completely different; however, by means of the fractional dynamical distributed controller, the systems synchronize (see Figs. 7.16 and 7.17). In Fig. 7.18 we observe how the synchronization error for activator and substrate concentrations along the space coordinate tends to zero. On the other hand, control signals are shown in Fig. 7.19. Notice how these present a low energy consumption.

In Fig. 7.20 we can observe how the gains affect the synchronization error. In Figs. 7.20a, b is shown how the synchronization error is unstable when the gains do not fulfill the stability criterion. On the other hand, the results in Figs. 7.20c, d demonstrate how small gains are capable to synchronize the PDE-based systems. It is obvious that the best results are obtained when the values of the gains are high, see Fig. 7.20e, f. Nonetheless, as the gains increased, the computational effort increases as well.

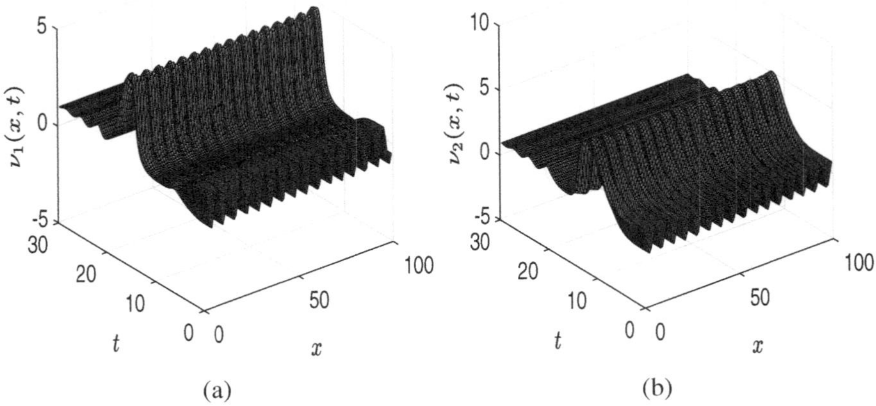

Fig. 7.14 Activator (**a**) and substrate (**b**) concentrations of the Schnakenberg master system in the original coordinates

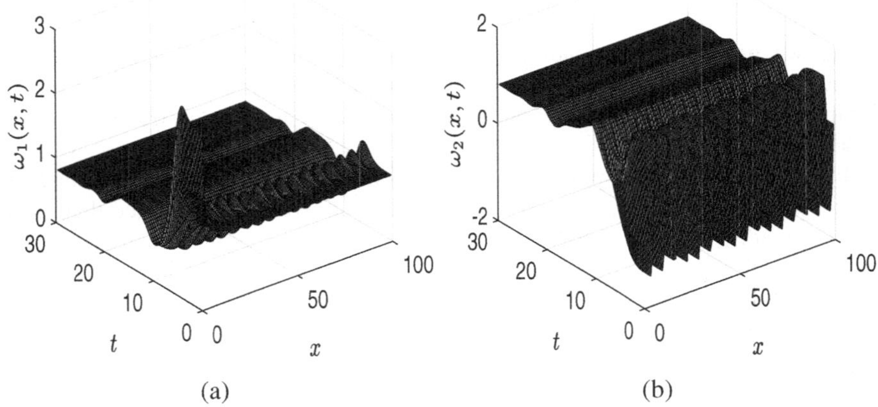

Fig. 7.15 Activator (**a**) and substrate (**b**) concentrations of the Schnakenberg slave system in the original coordinates

7.3.3 Heat and Moisture Concentration

Let us consider an advection-diffusion system given by two coupled PDEs of fractional order. This system describes heat and moisture transfer in a porous media, such as concrete materials. Many researchers have found that this diffusion process is of the anomalous kind. That is, particles diffuse slower than in a normal diffusion process and their features cannot be accurately described by traditional models. Therefore,

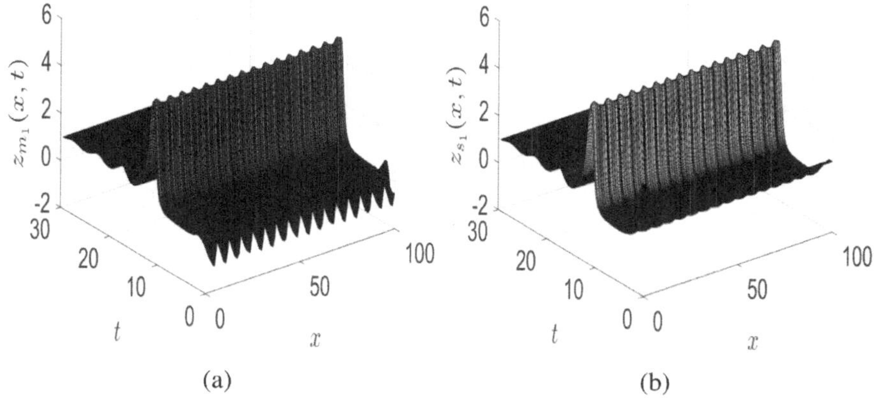

Fig. 7.16 Activator concentrations of the master system (**a**) and the slave system (**b**) in the transformed coordinates

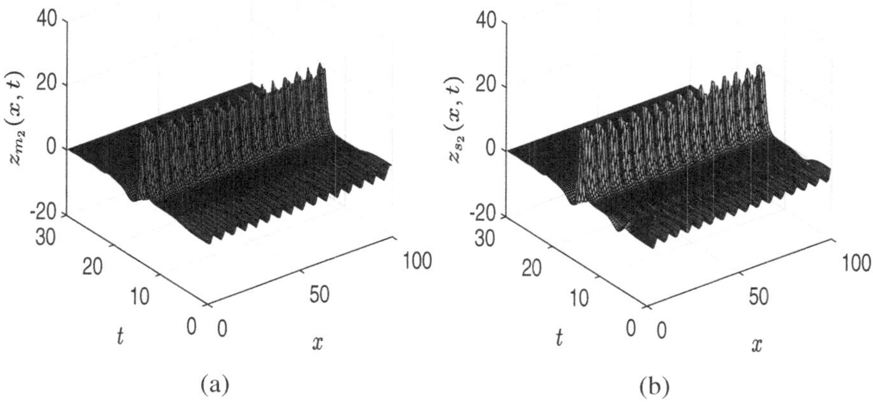

Fig. 7.17 Substrate concentrations of the master system (**a**) and the slave system (**b**) in the transformed coordinates

researchers have suggested using fractional-order models in order to fit better the experimental data [5, 43]. This problem is of high importance since the durability of reinforced concrete structures, particularly those in coastal areas, depends on moisture and corrosion.

Let us consider the following dimensionless advection-diffusion system of fractional order

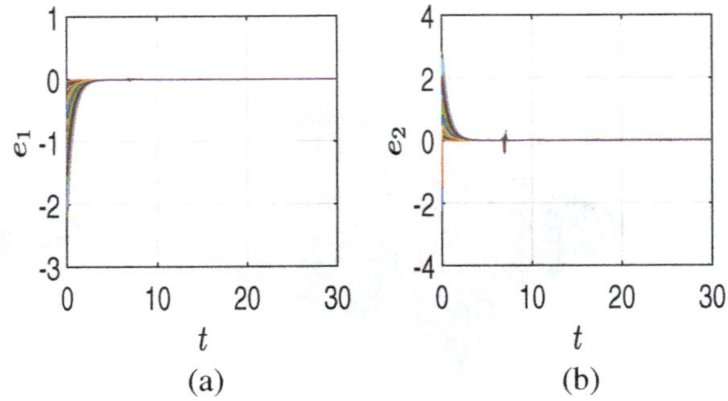

Fig. 7.18 Components of the synchronization error along the space coordinate **a** e_1 and **b** e_2

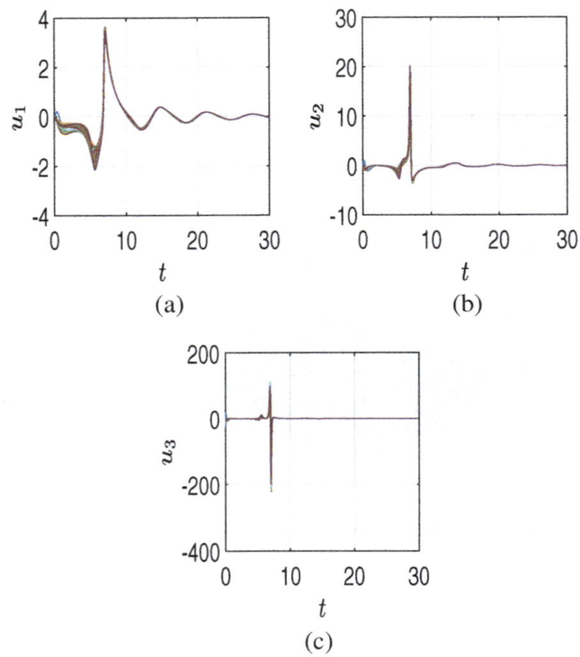

Fig. 7.19 **a** Fractional dynamical distributed control signals along the space coordinate and is **b** first and second **c** integrals

$$
\begin{aligned}
&\mathscr{D}^{(\alpha)} v_1(x,t) = -a_{11}\frac{\partial v_1(x,t)}{\partial x} + d_{11}\frac{\partial^2 v_1(x,t)}{\partial x^2},\\
&\mathscr{D}^{(\alpha)} v_2(x,t) = -a_{22}\frac{\partial v_2(x,t)}{\partial x} + d_{22}\frac{\partial^2 v_2(x,t)}{\partial x^2} - a_{21}\frac{\partial v_1(x,t)}{\partial x} + d_{21}\frac{\partial^2 v_1(x,t)}{\partial x^2},\\
&y = v_1(x,t).
\end{aligned}
\tag{7.54}
$$

where $v_1(x,t) = P/P^0$ and $v_2(x,t) = T/T^0$. Here, P and T are vapor pressure and temperature, respectively. The values P^0 and T^0 are environment reference

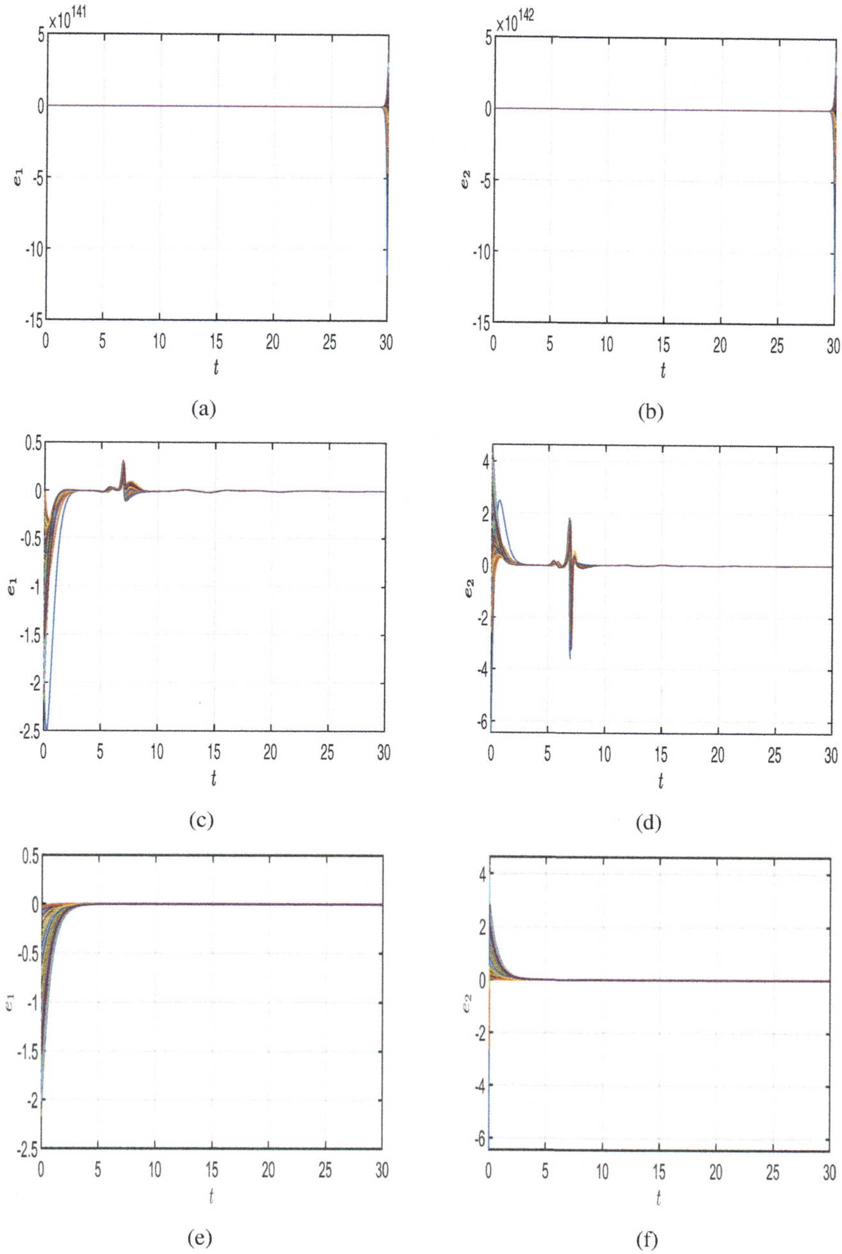

Fig. 7.20 Synchronization error with different gain values: $\kappa_1 = -10$ and $\kappa_2 = -10$ (**a**, **b**), $\kappa_1 = 7$ and $\kappa_2 = 5$ (**c**, **d**), $\kappa_1 = 700$ and $\kappa_2 = 500$ (**e**, **f**). Each color represents the synchronization error at an specific point of the discretized space

values. The constants a_{11}, a_{22}, a_{21}, d_{11},d_{22} and d_{21} are dimensionless parameters that describe the material properties (see [4] for further details). In this example, we will not consider a master system. Instead, we set a reference moisture profile. Let us consider a dry profile, for which we use the following equation: $r(x, t) = x^3/3 + 0.1$. Then, let us select for the slave system (7.54) the following change of variable: $z_1 = y + u_1$, such that we generate the following transformation:

$$\phi = \begin{pmatrix} z_1 \\ \mathscr{D}^{(\alpha)} z_1 \end{pmatrix} = \begin{pmatrix} y + u_1 \\ \mathscr{D}^{(\alpha)} y + u_2 \end{pmatrix} \tag{7.55}$$

Therefore, system (7.54) can be expressed as

$$\begin{aligned} \mathscr{D}^{(\alpha)} z_1 &= z_2, \\ \mathscr{D}^{(\alpha)} z_2 &= \mathscr{L}(z_1 + z_2) + \mathscr{D}^{(\alpha)} u_2, \end{aligned} \tag{7.56}$$

where $\mathscr{L}(z_1, z_2) = \mathscr{D}^{(2\alpha)} z_1$. Then, in this case we have that the synchronization error is $e = r - z$. Therefore, we have that

$$\begin{aligned} \mathscr{D}^{(\alpha)} e_1 &= e_2, \\ \mathscr{D}^{(\alpha)} e_2 &= \mathscr{D}^{(\alpha)} r - \mathscr{L}(z_1, z_2) - \mathscr{D}^{(\alpha)} u_2 \end{aligned} \tag{7.57}$$

In such a way that we propose $\mathscr{D}^{(\alpha)} u_2 = D^{(\alpha)} r - \mathscr{L}(z_1, z_2) + \kappa e$. Where $\kappa = [\kappa_1, \kappa_2]^T$. Therefore, the dynamical distributed controller is given by:

$$\begin{aligned} \mathscr{D}^{(\alpha)} e_1 &= e_2, \\ \mathscr{D}^{(\alpha)} e_2 &= \mathscr{D}^{(\alpha)} r - \mathscr{L}(z_1, z_2) - \mathscr{D}^{(\alpha)} u_2, \\ \mathscr{D}^{(\alpha)} u_1 &= u_2, \\ \mathscr{D}^{(\alpha)} u_2 &= \mathscr{D}^{(\alpha)} r - \mathscr{L}(z_1, z_2) + \kappa e \end{aligned} \tag{7.58}$$

which leads us to the following closed-loop system

$$\mathscr{D}^{(\alpha)} e = Ae, \tag{7.59}$$

with

$$A = \begin{bmatrix} 0 & 1 \\ -\kappa_1 & -\kappa_2 \end{bmatrix} \tag{7.60}$$

Then, the system is asymptotically stable if κ_1 and κ_2 are such that

$$|arg\left(-\frac{1}{2}\left(\kappa_2 \pm \sqrt{\kappa_2^2 - 4\kappa_1}\right)\right)| > \frac{\alpha\pi}{2}$$

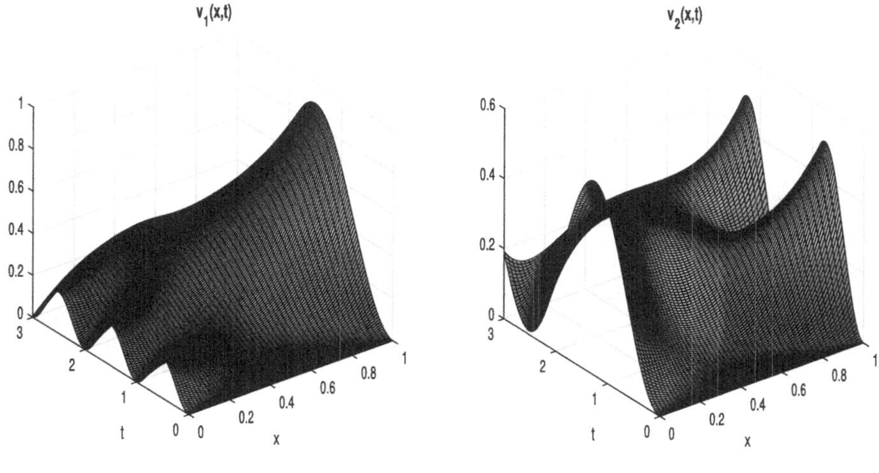

(a) (b)

Fig. 7.21 Original moisture (**a**) and heat (**b**) transfer behavior

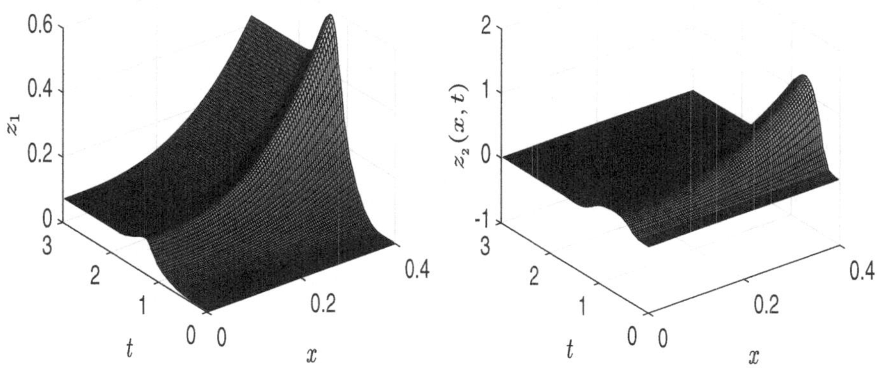

(a) (b)

Fig. 7.22 Moisture (**a**) and heat (**b**) transfer using the dynamical distributed controller. Notice how the desired moisture profile $r(x, t)$ is follow

For the numerical simulation, we consider the following parameters: $a_{11} = 0.02$, $a_{22} = 0.03$, $a_{21} = 0.01$, $d_{11} = 0.09$, $d_{22} = 0.07$ and $d_{21} = 0.03$. The initial conditions are as follows: $\nu_1(x, 0) = \nu_2(x, 0) = 0$. In Fig. 7.21 we can observe the original behavior of the system; on the other hand, in Fig. 7.22 we observe how the moisture profile reference is followed.

References

1. Aguila-Camacho, N., Duarte-Mermoud, M.A., Delgado-Aguilera, E.: Adaptive synchronization of fractional Lorenz systems using a reduced number of control signals and parameters. Chaos Solitons Fractals **87**, 1–11 (2016)
2. Artin, E.: The Gamma Function. Courier Dover Publications (2015)
3. Behinfaraz, R., Badamchizadeh, M.: Optimal synchronization of two different incommensurate fractional-order chaotic systems with fractional cost function. Complexity **21**(S1), 401–416 (2016)
4. Berger, J., Gasparin, S., Dutykh, D., Mendes, N.: On the solution of coupled heat and moisture transport in porous material. Transp. Porous Media **121**(3), 665–702 (2018)
5. Chen, W., Zhang, J., Zhang, J.: A variable-order time-fractional derivative model for chloride ions sub-diffusion in concrete structures. Fractional Calculus Appl. Anal. **16**(1), 76–92 (2013)
6. Cherif, M.H., Ziane, D.: Variational iteration method combined with new transform to solve fractional partial differential equations. Univ. J. Math. App. **1**(2), 113–120 (2018)
7. Delshad, S.S., Asheghan, M.M., Beheshti, M.H.: Synchronization of n-coupled incommensurate fractional-order chaotic systems with ring connection. Commun. Nonlinear Sci. Numer. Simul. **16**(9), 3815–3824 (2011)
8. Deng, W.: Generalized synchronization in fractional order systems. Phys. Rev. E **75**(5), 056201 (2007)
9. Dörfler, F., Bullo, F.: Synchronization in complex networks of phase oscillators: a survey. Automatica **50**(6), 1539–1564 (2014)
10. Gómez-Aguilar, J.: Space-time fractional diffusion equation using a derivative with nonsingular and regular kernel. Physica A: Stat. Mech. Appl. **465**, 562–572 (2017)
11. Hale, J.K.: Diffusive coupling, dissipation, and synchronization. J. Dyn. Differ. Eqs. **9**(1), 1–52 (1997)
12. Hemeda, A.: Variational iteration method for solving nonlinear coupled equations in 2-dimensional space in fluid mechanics. Int. J. Contemp. Math. Sci. **7**(37), 1839–1852 (2012)
13. Hristov, J.: Linear viscoelastic responses and constitutive equations in terms of fractional operators with non-singular kernels-pragmatic approach, memory kernel correspondence requirement and analyses. Eur. Phys. J. Plus **134**(6), 283 (2019)
14. Leibnitz, G.: Letter from Hanover, Germany, September 30, 1695 to ga l'Hôpital. Leibnizen Mathematische Schriften. Olms Verlag, Hildesheim, Germany (1962)
15. Lin, D., Wang, X.: Observer-based decentralized fuzzy neural sliding mode control for interconnected unknown chaotic systems via network structure adaptation. Fuzzy Sets Syst. **161**(15), 2066–2080 (2010)
16. Machado, J.T., Kiryakova, V., Mainardi, F.: Recent history of fractional calculus. Commun. Nonlinear Sci. Numer. Simul. **16**(3), 1140–1153 (2011)
17. Makris, N., Efthymiou, E.: Time-response functions of fractional-derivative rheological models. arXiv:2002.04581 (2020)
18. Martínez-Guerra, R., Cruz-Ancona, C.D., Pérez-Pinacho, C.A.: Generalized multi-synchronization viewed as a multi-agent leader-following consensus problem. Appl. Math. Comput. **282**, 226–236 (2016)
19. Martínez-Guerra, R., Mata-Machuca, J.L.: Fractional generalized synchronization in a class of nonlinear fractional order systems. Nonlinear Dyn. **77**(4), 1237–1244 (2014)
20. Matignon, D.: Stability results for fractional differential equations with applications to control processing. In: Computational Engineering in Systems Applications, vol. 2, Lille, France, pp. 963–968 (1996)
21. Momani, S., Odibat, Z.: Analytical approach to linear fractional partial differential equations arising in fluid mechanics. Phys. Lett. A **355**(4–5), 271–279 (2006)
22. Odibat, Z.M., Corson, N., Aziz-Alaoui, M., Bertelle, C.: Synchronization of chaotic fractional-order systems via linear control. Int. J. Bifurcation Chaos **20**(01), 81–97 (2010)
23. Oldham, K., Spanier, J.: The fractional calculus theory and applications of differentiation and integration to arbitrary order, vol. 111. Elsevier (1974)

24. Olfati-Saber, R., Fax, J.A., Murray, R.M.: Consensus and cooperation in networked multi-agent systems. Proc. IEEE **95**(1), 215–233 (2007)
25. Petráš, I.:Fractional-Order Nonlinear Systems: Modeling, Analysis and Simulation. Springer (2011)
26. Podlubny, I.: Fractional Differential Equations: An Introduction to Fractional Derivatives, Fractional Differential Equations, to Methods of Their Solution and Some of Their Applications, vol. 198. Elsevier (1998)
27. Podlubny, I.: Geometric and physical interpretation of fractional integration and fractional differentiation. arXiv:math/0110241 (2001)
28. Razminia, A., Baleanu, D.: Complete synchronization of commensurate fractional order chaotic systems using sliding mode control. Mechatronics **23**(7), 873–879 (2013)
29. Ren, W., Beard, R.W., Atkins, E.M.: Information consensus in multivehicle cooperative control. IEEE Control Syst Mag. **27**(2), 71–82 (2007)
30. Samko, S.G., Kilbas, A.A., Marichev, O.I., et al.: Fractional Integrals and Derivatives, vol. 1. Gordon and Breach Science Publishers, Yverdon Yverdon-les-Bains, Switzerland (1993)
31. Schnakenberg, J.: Simple chemical reaction systems with limit cycle behaviour. J. Theor. Biol. **81**(3), 389–400 (1979)
32. Strogatz, S.H.: Exploring complex networks. Nature **410**(6825), 268 (2001)
33. Tarasov, V.E. Fractional Dynamics: Applications of Fractional Calculus to Dynamics of Particles, Fields and Media. Springer (2011)
34. Tavazoei, M.S., Haeri, M.: Synchronization of chaotic fractional-order systems via active sliding mode controller. Physica A: Stat. Mech. Appl. **387**(1), 57–70 (2008)
35. Voyiadjis, G.Z., Sumelka, W.: Brain modelling in the framework of anisotropic hyperelasticity with time fractional damage evolution governed by the Caputo-Almeida fractional derivative. J. Mech. Behavior Biomed. Mater. **89**, 209–216 (2019)
36. Wang, J., Ma, Q., Chen, A., Liang, Z.: Pinning synchronization of fractional-order complex networks with Lipschitz-type nonlinear dynamics. ISA Trans. **57**, 111–116 (2015)
37. Wang, X., He, Y.: Projective synchronization of fractional order chaotic system based on linear separation. Phys. Lett. A **372**(4), 435–441 (2008)
38. Wang, X., Zhang, X., Ma, C.: Modified projective synchronization of fractional-order chaotic systems via active sliding mode control. Nonlinear Dyn. **69**(1–2), 511–517 (2012)
39. Wang, X.-Y., He, Y.-J., Wang, M.-J.: Chaos control of a fractional order modified coupled dynamos system. Nonlinear Anal.: Theory Methods Appl. **71**(12), 6126–6134 (2009)
40. Wang, X.-Y., Song, J.-M.: Synchronization of the fractional order hyperchaos Lorenz systems with activation feedback control. Commun. Nonlinear Sci. Numer. Simul. **14**(8), 3351–3357 (2009)
41. Wang, X.-Y., Wang, M.-J.: Dynamic analysis of the fractional-order Liu system and its synchronization. Chaos: Interdisci. J. Nonlinear Sci. **17**(3), 033106 (2007)
42. Wang, Y., Zhang, H., Wang, X., Yang, D. Networked synchronization control of coupled dynamic networks with time-varying delay. IEEE Trans. Syst. Man Cybern. Part B (Cybern.) **40**(6), 1468–1479 (2010)
43. Wei, S., Chen, W., Zhang, J.: Time-fractional derivative model for chloride ions sub-diffusion in reinforced concrete. Eur. J. Environ. Civil Eng. **21**(3), 319–331 (2017)
44. Wu, X., Lai, D., Lu, H.: Generalized synchronization of the fractional-order chaos in weighted complex dynamical networks with nonidentical nodes. Nonlinear Dyn. **69**(1–2), 667–683 (2012)
45. Wu, X., Li, J., Chen, G.: Chaos in the fractional order unified system and its synchronization. J. Franklin Inst. **345**(4), 392–401 (2008)
46. Yu, X., Zhang, Y., Sun, H., Zheng, C.: Time fractional derivative model with Mittag-Leffler function kernel for describing anomalous diffusion: analytical solution in bounded-domain and model comparison. Chaos Solitons Fractals **115**, 306–312 (2018)

Index

R. Martínez-Guerra and J. P. Flores-Flores, *An Approach to Multi-Agent Systems as a Generalized Multi-Synchronization Problem*, Understanding Complex Systems,
https://doi.org/10.1007/978-3-031-22669-4

GPSR Compliance

The European Union's (EU) General Product Safety Regulation (GPSR) is a set of rules that requires consumer products to be safe and our obligations to ensure this.

If you have any concerns about our products, you can contact us on ProductSafety@springernature.com

In case Publisher is established outside the EU, the EU authorized representative is:

Springer Nature Customer Service Center GmbH
Europaplatz 3
69115 Heidelberg, Germany

The manufacturer's authorised representative in the EU is Springer
Nature Customer Service Centre GmbH, Europaplatz 3, 69115 Heidelberg,
Germany. If you have any concerns regarding our products, please
contact ProductSafety@springernature.com

Printed and bound by CPI Group (UK) Ltd, Croydon, CR0 4YY
28/09/2026
02224087-0002